An Introduction to Liquid Scintillation Counting

An Introduction to Liquid Scintillation Counting

A. DYER

Department of Chemistry
University of Salford, U.K.

London · New York · Rheine

Heyden & Son Ltd., Spectrum House, Alderton Crescent, London NW4 3XX.

Heyden & Son Inc., 225 Park Avenue, New York, N.Y. 10017, U.S.A.

Heyden & Son GmbH, 4440 Rheine/Westf., Münsterstrasse 22, Germany.

ISBN 0 85501 092 4

Library of Congress Catalog Card No. 73–93364

Printed in Great Britain by J. W. Arrowsmith Ltd., Bristol BS3 2NT

To

Dilys, Andrew, Fiona,

Jennifer and Christopher

CONTENTS

PREFACE

This book is intended as an introduction to liquid scintillation counting for scientists and technicians with some knowledge of radiochemical technique. It will also serve as a source book for some more experienced practitioners and as a text-book for courses in radiochemistry, the medical use of radioisotopes, radiobiological assay and health physics at the undergraduate or post-graduate level. It does not attempt to provide answers to all the problems encountered in sample preparation for liquid scintillation counting. In the author's opinion, each problem is often specific to the user who should be able to choose the correct general method, and an accurate reference, from the information given in the relevant chapters.

My sincere thanks are due to Mrs. Barbara Ogundehin for her accurate interpretation of my untidy scrawl in preparing such an excellent typescript. Finally, my grateful thanks to my wife for constant support, cups of tea and, above all, keeping our children quiet whilst I was "otherwise occupied".

A. DYER

SALFORD

JANUARY 1974

ACKNOWLEDGEMENTS

The author gratefully acknowledges the permission, so freely given, from Dr. J. B. Birks (University of Manchester) to reproduce Figures 1 and 5 which are from his booklet "An Introduction to Liquid Scintillation Counting". This booklet was originally published jointly by Philips (Eindhoven) and Koch-Light Laboratories Ltd. Also to Libby Mueller for her kind agreement to the reproduction of a diagram (Figure 19) from her original publication (*Int. J. Appl. Radiat. Isotopes* **19**, 447 (1968)).

Figures 22 and 23 are used with permission of Grune and Stratton Inc. (N.Y.), as publishers of "The Current Status of Liquid Scintillation Counting" Editor E. D. Bransome Jnr. (1970), and Dr. Y. Kobayashi (The Worcester Foundation for Experimental Biology, Shrewsbury, Mass., U.S.A.) in whose article "Practical Aspects of Double-Isotope Counting" the diagrams originally appeared.

Finally Figure 26 is taken from an article by Dr. P. E. Stanley (Waite Agricultural Research Institute, The University of Adelaide, Glen Osmond, South Australia) who kindly gave his permission for its use as did the publishers, Academic Press Inc. (N.Y.), in whose book "Organic Scintillators and Liquid Scintillators" (Editors D. L. Horrocks and C-Z Peng, 1971) the article appears.

Chapter 1

THE TECHNIQUE

A. Introduction

The last twenty years have seen the growth of the technique of liquid scintillation counting to a point where it has almost replaced Geiger and proportional counting as methods for the detection of β radiation. It can also be a useful means of detecting α particles, positrons and weak X-rays.

This book attempts to provide an introduction to the use of the technique as well as describing briefly the physics of the liquid scintillation process. A contemporary bibliography of review articles and similar publications is provided at the end of the book whereby more detailed information, particularly of experimental methods, may be obtained. It is assumed that the reader has a working knowledge of the basis of nuclear chemistry.

B. The development of the technique

One of the earliest experiments in nuclear physics was the use of luminescent materials, by Rutherford, to detect the presence of α particles. His method relied upon the human eye to detect the flashes of light resulting from the impact of α particles on a zinc sulphide screen. This way of recording radiation was dormant until the development of the photomultiplier (PM) tube which is capable of detecting multiple light flashes (scintillations).

Liquid scintillation counting began in 1950 when Ageno, Kallman, Reynolds and their co-workers all reported that dilute solutions of certain organic substances could be used to measure β radiation.

Their main aims had been (a) to find scintillators with a high transparency to their own emitted light and (b) to improve the geometry of detection for electrons—particularly those of weak energy arising from the isotopes ^{14}C and ^{3}H (tritium). The major consequence of their work was a flexible technique whereby essentially 4π geometrical efficiency of detection was attained by intimately mixing the radio-isotope with an organic scintillator in a common solvent system. The method has the additional advantages of proportionality and ease of sample preparation with an almost complete absence of sample self-absorption.

In the early instruments the sample was placed in a suitable container (vial) on top of a PM tube. The impulses detected in the tube were linearly amplified and fed through a threshold discriminator (a single channel analyser) to a scaler and timing circuit. This simple instrumental arrangement has now evolved into a series of highly sophisticated instruments with multiple channel analysis, automatic sample handling and data outputs suitable for computer interfacing.

C. Scope of the technique

The largest use of the technique still lies in the measurement of ^{14}C and ^{3}H due to their wide applicability in biology, medicine, biochemistry and synthetic and mechanistic organic chemistry. The availability of multiple channel instruments enables the simultaneous determination of two (or even three) isotopes in the same sample provided that their β spectra can be sufficiently separated into distinct voltage channels. This certainly is the case for ^{14}C and ^{3}H.

There are other isotopes with comparatively weak β emissions and examples of these are in Table I. The table also includes isotopes whose positron emissions activate the scintillator, as well as isotopes of higher β energy all of which have been usefully measured by liquid scintillation methods.

Some isotopes decay via the emanation of a low energy X-ray. These are difficult to detect and in certain cases liquid scintillation is the only worthwhile method. The isotopes ^{55}Fe, ^{125}I and ^{51}Cr are of this type and are useful in nuclear medicine.

TABLE I. Some β^- and β^+ emitters which have been measured by liquid scintillation counting

Isotope	*Emission*	*Energy (MeV)*
^{3}H	β^-	0.018
^{63}Ni	β^-	0.067
^{129}I	β^-	0.15
	γ	0.038
^{14}C	β^-	0.155
^{35}S	β^-	0.167
^{45}Ca	β^-	0.254
^{59}Fe	β^-	0.27, 0.46
	γ	1.1, 1.3
^{131}I	β^-	0.335, 0.608
	γ	0.284, 0.364, 0.637
$^{137}Cs/^{137m}Ba$	β^-	0.514, 0.66, 1.20
^{90}Sr	β^-	0.545
^{36}Cl	β^-	0.714
^{40}K	β^-	1.33
	γ	1.46
	X-ray	0.0032
^{24}Na	β^-	1.39
	γ	1.37, 2.75
^{32}P	β^-	1.71
^{65}Zn	β^+	0.324
	γ	1.12
	X-ray	0.009
^{22}Na	β^+	0.54
	γ	0.51, 1.28

The use of organic scintillators in solution as detectors for α particles is amongst the earliest applications of the technique and provides a convenient way of calibrating α sources. When this factor is considered, and it is borne in mind that the Compton electrons associated with γ decay can also activate scintillators in solution, the flexibility and usefulness of liquid scintillation become apparent. Nuclear physicists have carried out absolute counting on many α and β emitters as well as performing studies on pulse height–

energy relationships for electrons and photoelectrons in liquid scintillation media. The method has also been used for half-life determinations.

One of the major areas of investigation into the application of liquid scintillation counting is that of sample preparation. The scientific literature contains hundreds of experimental methods designed to allow the incorporation of labelled samples into scintillation media. These samples are as diverse as radioactive rare gases (e.g. ^{85}Kr), blood, plasma and bone, cellular plant material and thin-layer chromatographic supports.

This introduction outlines the origins and scope of the method, and the following chapters provide a more detailed examination of the fundamentals and uses of liquid scintillation counting.

Chapter 2

FUNDAMENTALS

A. Some preliminary definitions

A scintillator may be defined as a material which emits a weak light flash (scintillation) when it interacts with certain quanta of radiation. The intensity of this scintillation is a function of the energy of the radiation deposited in the scintillator. When such a material is in solution it is described as a scintillator solute (although the terms phosphor and fluor also are used regularly). Some scintillation solutions contain two solutes, which then are designated as primary solutes and secondary solutes according to their relative functions.

The term primary solvent is used to imply (i) a substance whose function is to dissolve a solute and (ii) a substance capable of playing an important part in the energy transference sequence essential to the scintillation process. Secondary and ternary solvents may be needed to aid miscibility between a radioactive sample and a primary solvent–solute system. Mixtures of solutes and solvents are often referred to as "cocktails".

B. The series of events giving rise to a scintillation

The passage of ionizing radiation through a solvent–solute system causes ionization and excitation of the solvent molecules. The ionized molecules quickly recombine with electrons to form excited molecules. About 10% of these are the result of π electrons promoted from their ground state S_{0x} (Fig. 1) to states of higher energy described as singlet states. The remaining 90% of the excited solvent molecules are in σ electronic states which lose their energy thermally, thereby making no contribution to the scintillation process.

The excited π states rapidly undergo internal conversion to the lowest excited singlet state (S_{1x}) in about 10^{-13} s. Beyond this stage the molecule may lose its excitation energy by fluorescence or by the radiationless process of internal

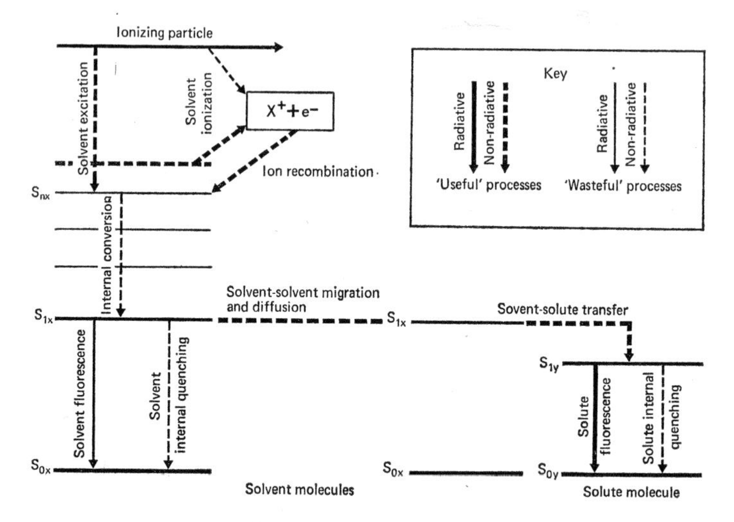

FIG. 1. The scintillation process (Birks[44]).

quenching. In practice the solvent molecules are migrating through the solution and most of them very quickly come into the environment of a solute molecule before they can lose energy. The probability of the solvent molecule losing energy by fluorescing is defined as the fluorescence quantum efficiency q, the ratio of the number of fluorescence photons produced per number of molecules originally excited. For toluene, a common primary solvent, q is about 0.1.

Once the excited solvent molecules are in proximity to solute molecules an energy transfer can take place as the lowest excited singlet state (S_{1y}) of the solute is lower than

S_{1x} (Fig. 1). If there is a sufficient concentration of solute present (~4 g/l) this energy transfer process has an efficiency of close to unity, i.e. nearly all the energy can be transferred from excited solvent to solute molecules. The final stage is a solute fluorescence representing the scintillation emission of the binary system, and this is another process of high quantum efficiency. This scintillation emission is that recorded by the PM. Fig. 2 is a schematic representation of the sequence of events producing a scintillation.

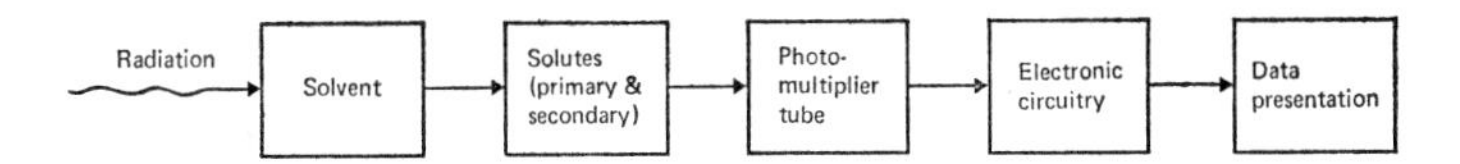

FIG. 2. Block diagram to illustrate the approximate experimental arrangement for a liquid scintillation determination.

C. The solvent

1. Solvent conversion factor

Generally the solvent will be of aromatic character thus providing π-bonded electrons capable of ready excitation by β radiation. A measure of this is given by the solvent conversion factor (s) defined by

$$A = sW,$$

where A is the number of excited solvent molecules raised to an S_{1x} state and W is the β energy. s corresponds to roughly one excited molecule per 100 eV of β energy. Toluene is used as a standard, $s = 100$, and for xylene $s = 107$, benzene $s = 85$ and anisole $s = 83$.

2. Energy transfer

The solvent patently must be able to pass energy to a solute with a high degree of efficiency. This is defined by the solvent–solute energy transfer quantum efficiency (f), which is dependent upon the concentration of the solute but is nearly independent of the nature of the solute. Hayes[17] carried out the first formal analysis of solvent properties for some 44 different liquids and found that benzene and the alkylbenzenes were the most suitable, a conclusion which is still held.

3. Secondary solvents

Attempts to introduce a wide range of samples into scintillation solutions often have involved the use of other solvents, either primarily to dissolve the sample or sometimes to provide mutual solubility for a solute and sample. The most common alternative primary solvent is 1,4-dioxane.

TABLE II. Volume of diluter causing a 25% reduction in scintillation efficiency when added to 10 ml toluene/PPO solution

Diluters	*Volume added (ml)*
o-Fluorotoluene	10
Fluorobenzene, mineral oil	6.5
Cyclohexane, *p*-fluorotoluene, methyl cyclohexane	5
Hexane, dicyclohexyl	6
Fluorocyclohexane, methyl borate	4.5
1,4-Dioxane	4
1,1-Dimethoxyethane	3.5
Cyclohexene	1.5
Methanol, ethanol, butanol, cyclohexanol	1
Thiophene	0.5

The presence of a secondary solvent may interfere with the energy transfer process, causing a drop in efficiency of measurement. Those materials which have a minor effect are described as diluters and Birks[45] has given a list of diluters which could be used as secondary or primary solvents. The limit of tolerance is arbitrarily designated as the quantity of diluter which can be added to a 10 ml scintillator solution whilst retaining 75% of the original scintillation efficiency (see Table II). Certain other materials cause more severe disruption to the sequence of events producing a scintillation and these are called quenchers. They present one of the major problems in experimental practice and are discussed on p. 57.

4. Solvent purity

The possibility of severe effects arising from the presence of small concentrations of quenchers means that special care must be taken to obtain solvents of high purity. One of the most common quenchers is dissolved oxygen, and experimenters requiring a high degree of accuracy use solvents from which oxygen has been removed by nitrogen bubbling. These "oxygen free" solvents are available commercially. Toluene is marketed as "scintillator grade" although "sulphur-free" toluene may be satisfactory for routine work.

Dioxane is liable to contain deleterious impurities (acetic acid, acetaldehyde, glycol acetate, and peroxides) and specially purified dioxane containing an antoxidant (0.01% butylated hydroxytoluene) is available. This must be stored under nitrogen and in the absence of light.

5. Other criteria

Advantageous properties for a solvent include a low vapour pressure, a low freezing point and a low optical density in the appropriate spectral range. This last point is not critical when the scintillator volume is small (5 to 20 ml) but must be remembered in the case of the large volume scintillators used in nuclear particle research and for the "whole-body" scintillators used in nuclear medicine. A summary of the requirements for a scintillator solvent is given in Table III.

TABLE III. General requirements for a liquid scintillation solvent

Available to a reasonable degree of purity
Utility as a solvent for samples to be measured
Ability to efficiently transfer excitation energy to a solute
A low optical density in the appropriate spectral range
Low vapour pressure
Low freezing point
Economic cost

6. Other solvents

Cocktails have been designed including a variety of solvents other than those already mentioned (Table VI, p. 30). They have included cumene, *p*-cymene, anisole, 2-ethoxyethanol, ethylene glycol, ethylene glycol monoethyl ether and carbitols. They have come into use in attempts to incorporate maximum amounts of aqueous samples. Naphthalene is often added to water-bearing cocktails in which it seems to perform as an additional solvent "bridging" the energy transference process between sample–solvent and solute. Recently benzo- and acetonitrile have been suggested[64] for use as solvents but these are as yet untried in current usage.

It is appropriate to mention that triethylbenzene, decalin, silicone oil and polyalkylbenzenes are recommended as solvents in large volume detectors because of their high flash points.

D. The primary solutes

1. Spectral properties

The solute is expected to have an emission range close to the response range of the PM tube. Similarly, lack of spectral overlap between emission and absorption spectra will be a desirable solute property. Hayes[69] used relative pulse height (r.p.h.) as a criterion for these properties as it can be shown that this is closely related to counting efficiency. However, this mode of analysis is a function of the particular instrumental arrangement and Birks[6,44] has suggested a more comprehensive comparison involving the definition of a spectral matching factor m. An accurate assessment of the correct comparison is beyond the scope of this book but the factor m will be used again and those readers who are concerned as to its exact meaning and origin should refer to the texts noted in the Bibliography.

2. Energy transfer

The efficiency f of solvent–solute energy transfer must be high and it has been demonstrated that transference is concentration dependent, maximizing at about 0.1 M (roughly 4 g/l). At this concentration f is close to unity. The solutes have fluorescence quantum efficiencies q close to one, where q is the ratio of fluorescence photons emitted to the number of solute molecules originally excited.

3. Other criteria

It is desirable that the solute is available in a pure form and that it be chemically and thermally stable. Reasonable

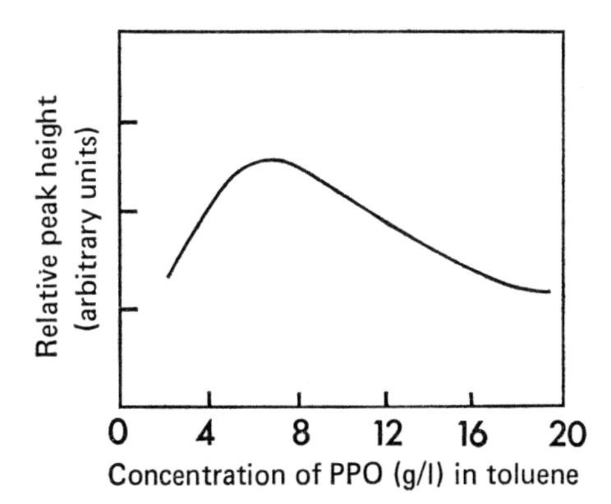

FIG. 3. Concentration quenching of PPO in toluene

solubility in solvents over an adequate temperature range and general compatibility, in solution, with varying sample types are obvious other requirements.

4. Quenching

Once a certain concentration of solute is exceeded then a drop in overall efficiency is observed (Fig. 3). This is called self-quenching and should be minimized. The resistance of a solute to quenching by other agencies is an important criterion for its use; in general, a solute with a short fluorescence decay time will be less liable to quenching. The general requirements for a scintillator solute are summarized in Table IV.

E. Survey of primary solutes (Fig. 4)

1. TP [*p*-terphenyl]

This was one of the original solutes and still compares favourably with those which have tended to replace it in that it has a high fluorescence quantum yield and its spectral matching factor to the bialkali PM tubes used in modern

TABLE IV. General requirements for a liquid scintillation solute

Emission range close to response range of photomultiplier
Lack of spectral overlap between emission and absorption spectra
High relative pulse height
Short scintillation decay time
Absence of self-quenching
Available to a high degree of purity
Efficient energy transfer participation
Resistance to quenching
Long "shelf-life"
Reasonable solubility in scintillator solvents
Compatibility with sample
Economic cost

instruments is good. However, it has a poor solubility in toluene, particularly at reduced temperatures, and it has a very limited compatibility with water.

2. PPO [2,5-diphenyloxazole]

PPO was proposed as a replacement for TP on the grounds of its high solubility in toluene, its compatibility with aqueous samples, and with low temperature operation. It is the most widely used solute.

3. PBD [2-phenyl-5-(4-biphenyl)—1,3,4-oxadiazole]

PBD is the most efficient primary solute in terms of spectral and energy transfer properties. However, it has a limited solubility and also requires about twice the concentration used for PPO to realize its greater efficiency.

4. Butyl-PBD(2-(4′-*t*-butylphenyl)-5-(4″-biphenyl)-1,3,4-oxadiazole)

This solute has fluorescence properties comparable to TP and PBD; it has excellent solubility in scintillation solvents. It also has good sample compatibility and is resistant to quenching. Its use is increasing.

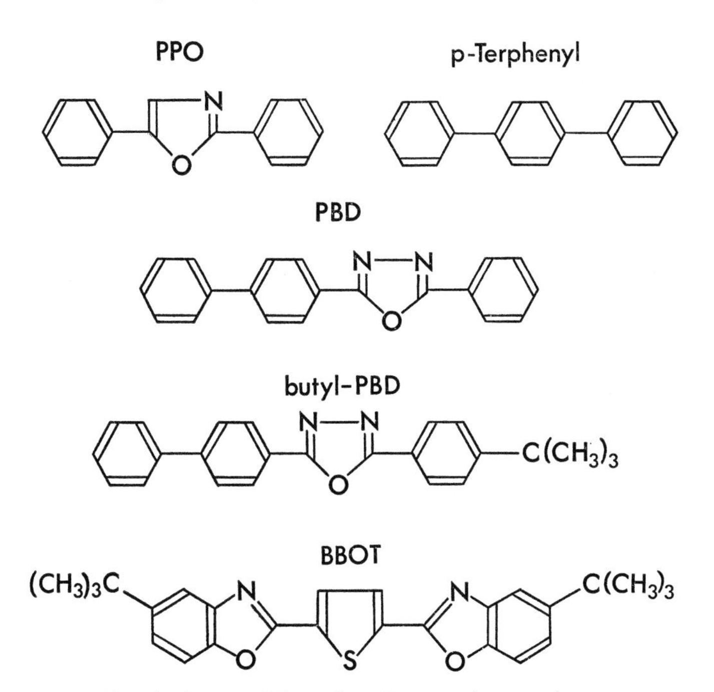

FIG. 4. Structural formulae of some primary solutes.

5. BBOT [2,5-bis-(5′-*t*-butylbenzoxazolyl(2′))-thiophene]

BBOT has an unusually long wavelength emission at 435 mμ, and in other properties it is generally comparable to PPO. It has been the subject of a critical assessment by Bush and Hansen,[53] who found it to be less efficient than PPO and to be much less soluble in the presence of water.

6. Other primary solutes

Many other chemicals have been suggested as liquid scintillation solutes: several are substituted oxazoles (e.g. PBO, 2-phenyl-5-(4-biphenyl)oxazole), and others are *p*-oligophenylenes (e.g. BIBUQ—a butylated quaterphenyl). None of these alternatives has been adopted for general use.

F. The design of new solutes

The major aim in the synthesis of a new solute is to reduce its liability to quenching but retain good fluorescence properties. This can be approached in two ways, (i) by destroying the ability of the chromophore to retain a coplanar configuration relative to the rest of the solute molecule and (ii) by shielding the chromophore with a bulky substituent group. Both these approaches try to prevent close contact between the solute and quencher.

A passing reference can be made to attempts to produce a substance combining both solute and solvent properties. The material l-methylnaphthalene has these desirable properties but is not compatible with water and has not been available commercially in sufficient purity until comparatively recently.

G. Scintillator solution figure of merit

Birks[44] has presented a useful concept for the assessment of combined solvent–primary solute efficacy by the definition of the scintillator solution figure of merit F as

$$F = sfqm$$

where s is the solvent conversion energy, f is the solvent–solute energy transfer efficiency, q is the fluorescence quantum efficiency of the solute and m is the spectral matching factor. F is a function of the PM and for a bialkali photomultiplier instrument 3 g/l PPO in toluene has $F = 73$ and 10 g/l PBD in xylene has $F = 81$.

H. The secondary solute

Secondary solutes have been included in cocktails to obtain a better match between the scintillation emission and the response range of the PM tube. Small concentrations (~0.4 g/l) of other fluorescent solutes may be added to create a further, high efficiency, energy transfer stage between the primary solute and this additional secondary solute.

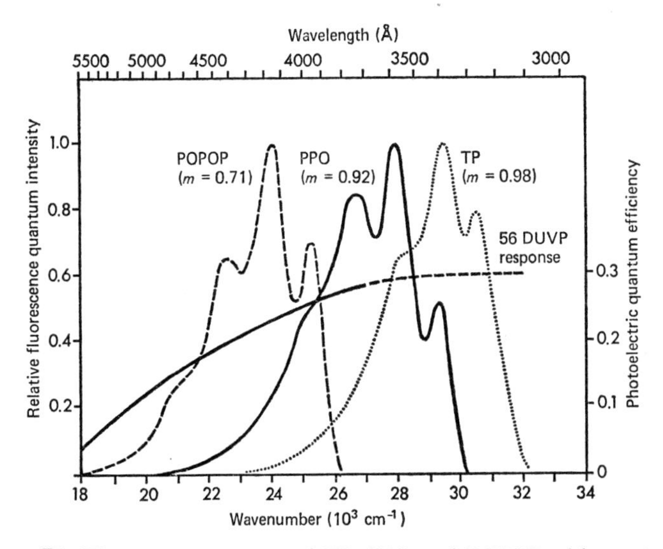

FIG. 5. Fluorescence spectra of TP, PPO and POPOP with spectra response of a bialkali PM. (m = spectral matching factor).[44]

The final scintillation principally is that from the return of excited secondary solute molecules to their ground state. These secondary solutes were chosen so that their mean fluorescence wavelength was more in accord with the caesium + antimony photocathodes of the PM tubes in the earlier instruments (hence their description as "wavelength–shifters" in earlier publications).

Modern instruments contain bialkali PM tubes which have a wider spectral response range which should eliminate the need for a secondary solute (Fig. 5). Despite this fact several workers have shown that, under certain conditions, it can still be an advantage to include a secondary solute in a liquid scintillation cocktail. These conditions are when either (i) the sample to be counted is coloured or opaque or (ii) the volume of the scintillator being used is large enough to cause the problem of self-absorption of the primary solute fluorescence. In regard to (i) it should be noted that a visual inspection for colour can be misleading and a check on the absorption spectrum is advisable.

The quantum efficiency of the transfer of energy between primary and secondary solutes is about 0.5 to 1.0 and depends upon the overlap between the fluorescence spectra of the primary solute and the absorption spectra of the secondary solute. Birks has expressed the criteria for the use of a secondary solute in terms of q and m such that, in the absence of colour, a wavelength shifter is only an advantage when $q'm' > qm$. (q,q' and m,m' are the fluorescence quantum efficiencies and the spectral matching factors of the primary and secondary solutes respectively).

J. Some secondary solutes (Fig. 6)

1. POPOP [1,4-di-(2-(5-phenyloxazolyl))benzene]

This is the most widely used secondary solute and is compatible with most of the primary solutes, solvents and samples commonly encountered.

2. DMPOPOP [1,4-di-(2-(4-methyl-5-phenyloxazolyl)) benzene]

The dimethyl derivative of POPOP suggested as an alternative to POPOP due to its greater solubility in toluene but it needs to be used in a higher concentration to achieve an equivalent efficiency.

3. bis-MSB [*p*-bis-(σ-methylstyryl)benzene][54]

This solute has a good solubility in toluene and is less liable to quench effects than both POPOP and DMPOP. A

POPOP

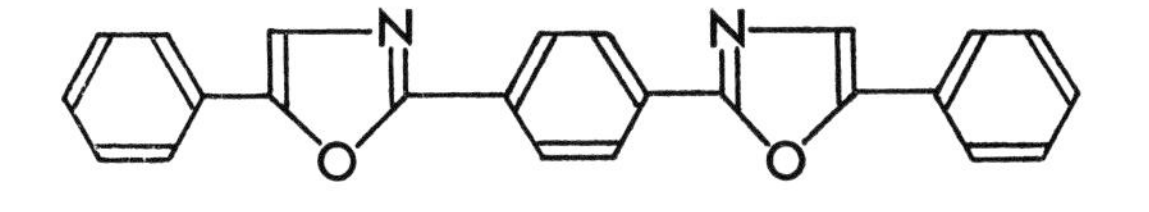

Bis-MSB

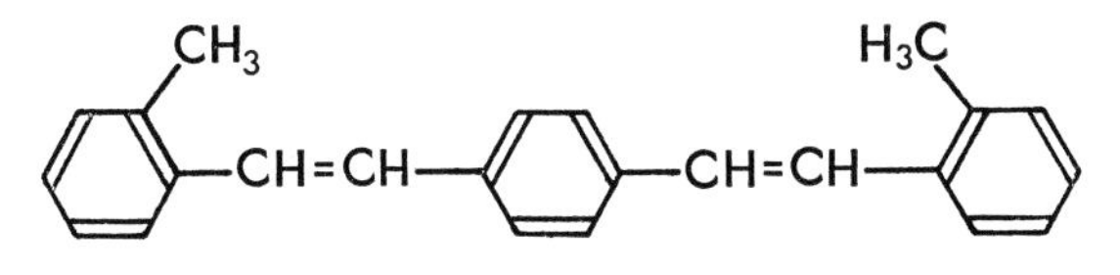

BBO

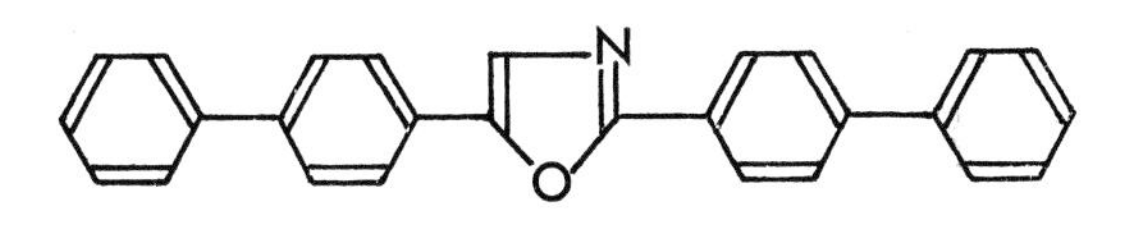

dimethyl-POPOP

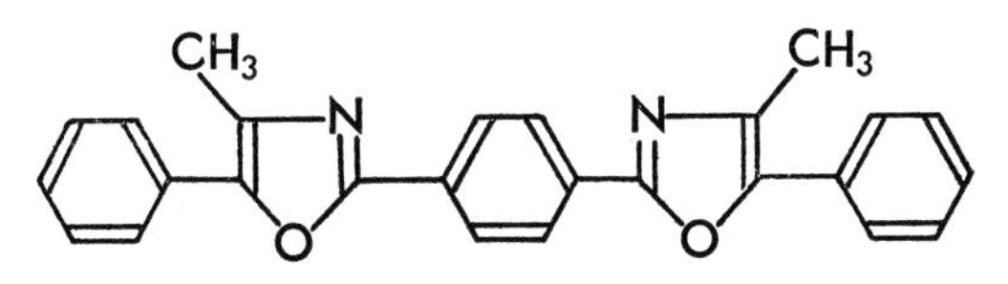

α-NPO

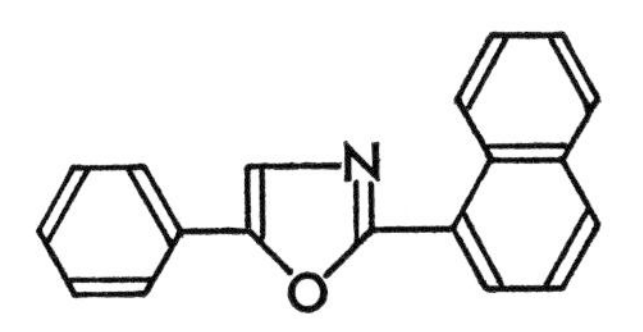

DPHT

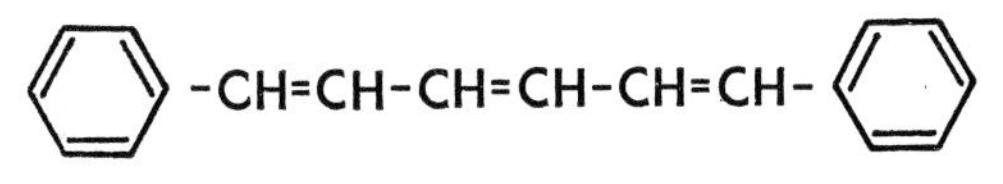

FIG. 6. Structural formulae of some secondary solutes.

similar substance is bis-PSB[1,4-bis-(4-isopropylstyryl)benzene] which must be used in higher concentrations and probably has no advantage in its use.

4. PBBO [2-(4-biphenyl)-6-phenylbenzoxazole]

PBBO is recommended for use in large volume counters.

5. Other secondary solutes

Again many compounds have been tried as secondary solutes. Those which occur with some frequency in the literature are: α-NPO [2-(1-naphthyl)-5-phenyloxazole], BBO [2,5-di-(4-biphenyloxazole)] and the original wavelength shifter associated with TP, namely DPHT [1,6-diphenyl-1,3,5-hexatriene].

Chapter 3

INSTRUMENTATION

A. Early equipment

The earliest instruments were adaptions of single PM tube γ counters linked to a single channel analyser via a preamplifier and amplifier. The sample container usually was on the face of a vertically positioned PM tube. Optical coupling was by silicone oil, and a shutter device protected the PM tube from exposure to daylight. Reflectors of aluminium or titanium dioxide paint were positioned to maxi-

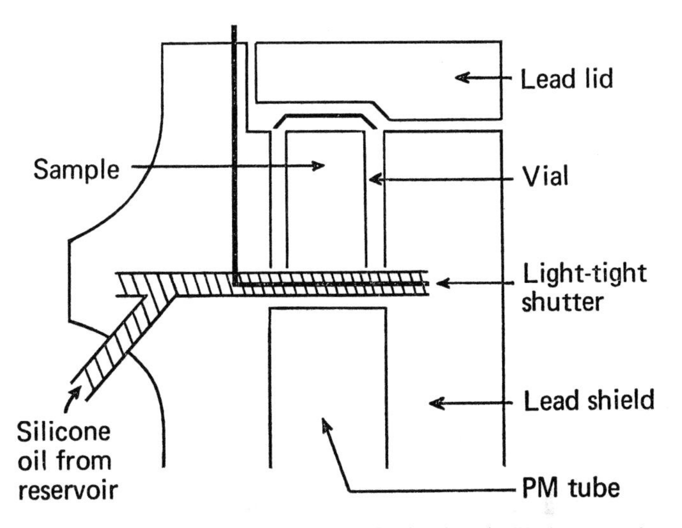

FIG. 7. A single PM tube arrangement for liquid scintillation counting.

mize light collection at the photocathode. This simple experimental arrangement is shown in Fig. 7. These early machines were cooled, either by water or refrigeration circuits, to minimize thermal noise on the PM tube.

B. Development of the modern instrument[9]

The first commercial instruments were marketed by Packard in 1953 and adopted the feature of coincidence counting developed a year or so earlier by Heibert and Hayes at Los Alamos. This involved the use of two horizontally

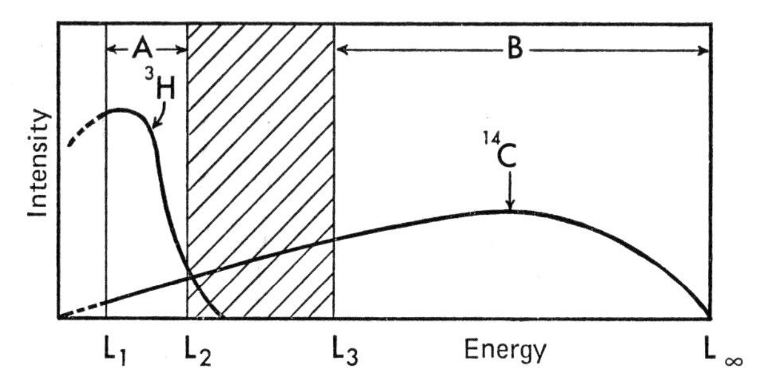

FIG. 8. Schematic diagram of A and B channels arranged to count ^{3}H and ^{14}C in two counting windows. Windows are defined by discriminator levels $L_1 \rightarrow L_2$ and $L_3 \rightarrow L_\infty$. Illustrates relatively poor energy resolution of early counters operated at 100% gain (ie. without attenuation).

mounted PM tubes to view a sample and only events recorded as occurring simultaneously by the two tubes were retained for counting. All "non-coincident" events were presumed to be noise. The resolving time was about 1 μs. An additional advantage of this arrangement was that the tubes could be run at high voltages and hence high gains without the occurrence of prohibitively high backgrounds. The arrangement had the inherent disadvantage that the light output from each scintillation was recorded as two portions (from each of the photomultipliers) with the consequent reduction of signal amplitude. 1957 saw the advent of 100 sample capacity instruments with electronic timing and data print-out facilities. Transistorization became universal in the early 1960's.

In 1961 a spectrometer was manufactured with two distinct voltage channels each having its own attenuated fixed gain amplifier. It was a two tube instrument with the output from one tube subjected to pulse-height analysis and the other tube functioning as a monitor to check coincidence. The first of these instruments was the Packard 314E and its circuitry enabled the simultaneous determination of two isotopes at the so-called "balance-point". Previous two-channel instruments operated at an unsatisfactory resolution

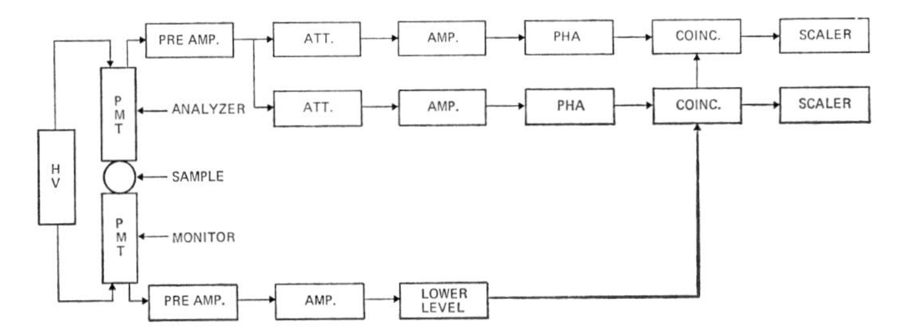

FIG. 9. Packard 314E series (1961), improved two channel system logic (Rapkin, ref. 9).

of β energies (Fig. 8). The effect of "balance point" counting was to centralize the individual spectrum in the selected counting "window" (i.e. between two discriminator voltages). A block plan of the circuit used is shown in Fig. 9.

The next generation of instruments was much improved by the use of the continuous chain sample belt with the added possibilities of by-passing groups of samples. This advance was made in 1962 by Nuclear-Chicago in an instrument which also provided three counting channels to combat quenching problems using a "channels-ratio" method. This method, and other procedures for dealing with quenching, are considered in Chapter 6, p. 60). This instrument also had, as an optional extra, a mechanical calculator to assist the data-handling produced by channels-ratio.

An important change in instrument logic came about in 1963 when Packard designed a series of products in which the pulses from the PM tubes were summed rather than divided. This improved both the amplitude and resolution of the pulses. This major advance now forms part of the design of all modern automatic liquid scintillation spectrometers. An indication of the benefit from this change is in the improvement in the efficiency of tritium determinations from roughly 30 to 45% with a reduction in backgrounds from 50 to 30 counts/min. The facilities for "background subtraction" and "low-count reject" as part of the automatic operation first appeared on commercial instruments in this year.

An ANS itron counter was marketed in 1964 introducing three new design concepts: first, quench correction was by the use of an external standard γ source; second, logarithmic amplifiers were used to eliminate the need for attenuation controls; third, preadjusted counting windows were provided. Although this instrument was not a commercial success all three features have been incorporated into current designs but are not used universally.

In 1965 Beckman produced their first instrument which was the first commercially successful machine to use preadjusted counting windows (via "plug-in" modules). This counter, and the contemporary Nuclear-Chicago product, also provided two additional counting channels reserved for external standardization (see p. 62). About this time integrated circuitry and bialkali PM tubes were being used. The improvements gained from bialkali tubes rested in their low noise levels, high efficiencies, and wider spectral response ranges. Both these innovations are used in most modern machines. The use of bialkali tubes should allow counting to be carried out at stabilized ambient temperatures and indeed ambient counters are available commercially. Despite this there are still advantages to be gained at reduced temperatures and most manufacturers preserve the cooled counting compartment.

TABLE V. Features of contemporary instrument design

Feature	*Comment*
Multi-user instruments	Each user has his own instrumental settings which are called up by a programmed code fed into the circuit
Continuous display of energy spectrum	Enables best location of the observed β spectrum in the optimum discriminator levels
Automatic decay correction	Programmable unit provided to compensate for the decay characteristics of the isotopes being measured
Colour restoration	Automatic compensation of deleterious effects caused by coloured samples
Cross-talk discrimination	Reduces errors caused by events due to background radiation occurring in one PM tube being recorded by the other (cross-talk)
Photon monitor	Distinguishes between true sample counts and single photon events occuring in a counting bottle caused by chemiluminescence, phosphorescence, and luminescence of chemical or biological origin
Automatic calibration	Ensures stability and reproducibility of counts by constant recalibration against a known standard
Combined β/γ counters	Optimizes the utility of one instrument in a "multi-isotope" laboratory
Automatic combustion liquid scintillation counters	Intractable samples are combusted automatically to $^{14}CO_2$ or THO which are then counted by trapping into a scintillator solution
Sample capacity	Racks, trays or continuous belts capable of handling up to 420 samples at one loading

C. Modern liquid scintillation spectrometers

At this time more manufacturers are competing in the liquid scintillation market. The advances in design have been varied and in concluding this chapter the most important advances have been summarized in Table V. It does not include the developments in data-handling which have progressed from the use, in 1966, of an interfaced electric typewriter to the present day options of programmable computation facilities or even an "on-line" link to a computer.

The amount of sophistication required in processing results from a liquid scintillation counter obviously is a matter for the individual user to decide and the decision to embark on methods involving computers should not be taken without due consideration of the costs involved. As a generalization only those users involved in a very high sample throughput, particularly of quenched samples and for double labelled samples, will be able to justify the use of a computer unless the use of an existing facility is available.

Chapter 4

SAMPLE PREPARATION FOR SUBSTANCES LABELLED WITH CARBON-14 AND TRITIUM

A. Homogeneous sample preparation

1. Samples soluble in a scintillator solvent

The happiest circumstance for a user of the liquid scintillation technique is that his labelled sample has good solubility in one of the standard solvents. This may be the case for lipids, steroids and simple aliphatic and aromatic compounds. Chemical modification to promote solubility may be worthwhile although this obviously is a function of the amount of modification required. The conversion of glucose to its penta-acetate is an example of a simple chemical conversion which is justified in terms of sample preparation. Samples directly soluble in a scintillator solvent may still be a problem to count when coloured solutions are produced. The approach to coloured solutions will be covered in later discussions.

2. The use of other solvents to achieve homogeneous samples

(a) TRITIATED WATER

The incorporation of aqueous samples into a scintillator solution was an early problem to be faced. The general principle was to use a blending solvent (diluter) to obtain miscibility between water and a solution of PPO and POPOP in toluene. The first blending solvent to be used was ethyl alcohol but its utility was limited by the maximum value of only 3% water capable of mutual miscibility in toluene + ethanol mixtures.

An alternative is to use a solvent which combines the advantages of water miscibility with the ability to hold a solute in a satisfactory concentration (~4 g/1). Such a solvent is 1,4-dioxane and this material has been a frequent constituent of cocktails designed to accept water and aqueous solutions. An important drawback to its use is its freezing point (11.7 °C) which precluded its use in the refrigerated counting compartments needed to minimize PM thermal noise in the earlier instruments. This may be avoided by adding ethylene glycol as an "antifreeze".

Dioxane is not an aromatic solvent and its quantum efficiency is low. This can be improved by additions of naphthalene (100 to 150 g/l) which restores some of the scintillator efficiency whilst retaining the capacity for aqueous samples of up to 20% of the total volume. Naphthalene functions as an intermediate secondary solvent. The use of dioxane, although still prevalent, should be viewed objectively. It will be appreciated that dioxane based scintillators are of necessity complex, although the addition of an "antifreeze" should not be required in a modern ambient counter. Dioxane is also expensive and liable to form peroxides which promote unwanted chemiluminescent reactions. Certainly, if dioxane is used, it should be stored in the dark and the addition of granulated zinc, or another antoxidant, to scavenge peroxides is essential (see also p. 77).

(b) OTHER AQUEOUS SOLUTIONS

Additional solvents which have been used, either to replace dioxane or in conjunction with it, to improve capacity for water and aqueous solutions are various methyl and ethyl ethers of ethylene glycol, "Cellusolve" and anisole. They are often used in conjunction with a xylene based scintillator solution.

A variety of aqueous samples of biological origin have been counted in dioxane cocktails and the literature reports the determination of proteins, carbohydrates, blood, urine and plasma in this way. The presence of appreciable concentrations of inorganic salts and buffer solutions may also complicate sample preparation.

There are several commercial products designed to incorporate water into cocktails. Their exact composition is not available and before deciding to use one of these products, or one of the traditional dioxane solutions, a potential user is recommended to refer to the use of the emulsion technique which is gaining in popularity when water has to be taken up into a cocktail. This method is discussed on p. 31.

(c) FIGURE OF MERIT

Kinard[80] compared systems for counting aqueous solutions by using a figure of merit defined as the product of the volume of aqueous sample taken into a cocktail multiplied by the observed counting efficiency. A high figure of merit will give an indication of a short counting time. As the figure of merit is a function of the individual instrument used, direct comparisons between results obtained in different laboratories are not advised except in the most general terms. (Note that this is not the same parameter as that defined on p. 33.)

3. The preparation of homogeneous samples by using solubilizers

Biological samples, such as tissues and plasma, were first taken into suitable solution by digestion with about 5 M alkali (NaOH or KOH) in alcoholic solution. This solubilization technique is unsatisfactory for tritium estimations and the homogeneity of the final solutions has been questioned.

(a) HYAMINE

Various alternative solubilizers have been investigated of which Hyamine 10 X [*p*- (di-isobutylcresoxethoxethyldimethylbenzylammonium hydroxide)] was the first commercially viable product. It can be used in small quantities with toluene or dioxane based solutions and can render soluble serum, plasma, amino acids, proteins, tissues, purines, pyrimidines, and sugars. It also takes into solution the polyacrylamide gels from protein fractionation.

Hyamine[40] has the disadvantages of producing chemiluminescent reactions particularly with basic substances and dioxane. It yellows during solubilization and is a quenching substance. Hyamine also is expensive. Often the disadvantage of chemiluminescence can be controlled by working in acid solution or by working at reduced temperatures. For low activity samples it may be advisable to store in the dark (without acid treatment) to allow the chemiluminescence to decay before counting. The presence of Hyamine exacerbates peroxide formation, and intense light production is noticeable in samples which have been bleached by the hydrogen peroxide or benzoyl peroxide treatment used to reduce colour in biological samples (e.g. blood and urine). The chloride form of Hyamine[9] has been used as an alternative and is said to give higher counting efficiencies for small sample sizes.

An early alternative to Hyamine was a similar material called Primene but its use has not continued. More recently several commercial solubilizers have been marketed. Soluene 100 and NCS are of the same chemical type as Hyamine, consequently suffering from the same general disadvantages. In a comparison between ethanolic KOH, Hyamine and NCS Bush and Hansen[66] show that NCS has a generally superior performance. NCS has been recommended for the counting of polyacrylamide gels.

Recently "Bio-solv" and "Unisolv" have become available. They offer quicker digestion times and freedom from chemiluminescence whilst retaining the ability to take up water in reasonable quantities. NCS, Soluene, and another product Protosol are reported to cause yellowing with butyl and isopropyl PBD.

A summary of the foregoing comments would be to avoid the use of a dioxane based solubilizer/scintillator in favour of a toluene scintillator if this is practicable. Nevertheless it must be remembered that, when water is present, the use of toluene may cause colloid formation. This may not be a deterrent to using the particular solubilizer mix but special care must be taken as will become apparent from subsequent discussions on emulsion counting (p. 31). Fox has examined this problem in detail and appealed to the manufacturers to avoid "secret recipes" because of the strong possibility of colloid formation which is difficult to detect and can cause erroneous results when quenching is being corrected for (see pp. 31, 58, 62).

A selection of cocktails relevant to the incorporation of water bearing and other biological samples is presented in Table VI and other useful information is tabulated in the Appendix, p. 103.

B. Heterogeneous sample preparation

1. Gel counting

One early method of counting large volumes of aqueous solutions was to allow an emulsion to form in a toluene based solution and then to add a substance to stabilize the emulsion. Radin successfully used a combination of glycerol and Thixcin (a castor-oil derivative) to count aqueous solutions containing ^{14}C compounds. An efficiency of 45 % was observed in an emulsion containing 2 % of water. This was not satisfactory for ^{3}H samples.

Later the fumed silica product Cab-O-Sil replaced Thixcin and was shown to provide stable emulsions with aqueous solutions and milk. It has a maximum water capacity of 6 % in dioxane + xylene + naphthalene scintillant. Cab-O-Sil is thixotropic and has found wide use in dealing with tissue homogenates; it creates samples which are patently inhomogeneous but which give reproducible and stable count rates with reasonable counting efficiencies. The gels formed by Cab-O-Sil have a limited water tolerance and it is probably true to say that most workers prefer to use a solubilization technique for tissue work whilst acknowledging that gels can give acceptable results.

Cab-O-Sil is a very convenient agent for the direct radioassay of thin layer chromatograms. Spots are scraped carefully from the chromatographic plate and dispersed directly into a toluene based scintillator with 4 % (w/v) of Cab-O-Sil. Most of the common thin-layer adsorbents have no deleterious effects on counting efficiencies but residual solvents may cause spurious results due to their quenching properties.

TABLE VI. Some cocktails for the incorporation of water and aqueous solutions into liquid scintillation media

Composition and Originator	*Comment*†
5 parts toluene containing 4 g/l PPO + 40 mg/l POPOP 4 parts ethanol Hayes and Gould[68]	Low total water capacity ~ 3% with ^{3}H efficiency ~ 2.8%
5 parts xylene 5 parts dioxane 3 parts ethanol 5 g/l PPO + 50 mg/l αNPO 80 g/l naphthalene Kinard[80]	Takes up ~ 7.7% water; ^{3}H efficiency ~ 4%
6 parts dioxane 1 part anisole 1 part 1,2-dimethoxyethane 5 g/l PPO + 50 mg/l POPOP Davidson DAM 611[10]	About 25% water capacity giving ^{3}H efficiency of 2.3% *or* 20% water with 4.3% ^{3}H efficiency
60 g naphthalene 100 ml methanol 20 ml ethylene glycol 4 g PPO + 200 mg POPOP (or DMPOPOP) make up to 1 litre with dioxane Bray[51]	Widely used, good for salt solutions and many biological samples. About 30% max. water capacity with 22% efficiency for ^{3}H
9 parts dioxane 1 part ethylene glycol monomethyl ether 50 g/l naphthalene 10 g/l PPO + 50 mg/l POPOP Polesky and Seligson[89]	About 30% water capacity with 28% ^{3}H efficiency

† Water content depends on temperature of counting—these values are at about − 5 °C. Efficiency is a function of instrument used; hence these are intended as a rough comparison.

Other labelled particulate materials, e.g. bone and barium [^{14}C]carbonate have been assayed adequately by similar techniques.

The success of the method derives from the absence of quenching effects as the sample is not in true solution with the scintillator solvent. It follows that samples should be checked to ensure that toluene (or dioxane) soluble substances are not being leached from the particulate material into the solvent. Obviously this causes unstable counting conditions due to quenching effects and/or varying counting efficiencies. Cab-O-Sil suspensions which are coloured should be avoided. Another use of Cab-O-Sil is to prevent the adsorption of trace quantities of labelled material on to the walls of glass counting bottles.

Other gel methods have been described[11] whereby a gel is created by reacting a toluene scintillator solution containing branched aliphatic primary amines (Armeen L-11) with bitolylene di-isocyanate (Isonate 136-T). Another method[7] is to heat a toluene scintillator with pellets of polyolefin resins (e.g. Poly-gel B). This latter technique can be used to incorporate barium carbonate, rat-liver powder, powdered plant material and thin-layer adsorbents. Poly-gel B is not compatible with aqueous solutions but it does not quench and has a high transmittivity to light.

NOTE. When counting solids suspended in a gel the particle size or size distributions must be controlled carefully. This may involve grinding and sieving, which are time-consuming, although precipitation under controlled pH conditions from homogeneous solution may be an alternative. If particle size is not controlled then unknown counting losses due to the self-absorption of β particles in the solid can occur.

2. Emulsion (colloid) counting

Patterson and Greene,[87] in examining some earlier work by Meade and Stiglitz,[84] were the first workers to appreciate that detergents had a broad application as toluene emulsifiers. As early as 1957 Radin[43] had used "Tween" and "Span" detergents but these created unstable emulsions. Meade and Stiglitz[84] had added Triton X-100 (iso-octyl phenoxy-polyethoxyethanol) to a mixture of bodyfluids in Hyamine 10-X. Patterson and Greene[87] extended the use of this cheap non-ionic surfactant and found that a 2 : 1 toluene + Triton mixture with 4 g/l PPO and 100 mg/l POPOP had a water capacity of 23 % and gave a figure of merit of 230 for ^3H.

This is a complex mixture, and both Benson[50] and Van der Laarse[94] cautioned that a very accurate knowledge of sample preparation, inhomogeneity, and temperature fluctuations was essential for the accurate use of the Triton X-100 method.

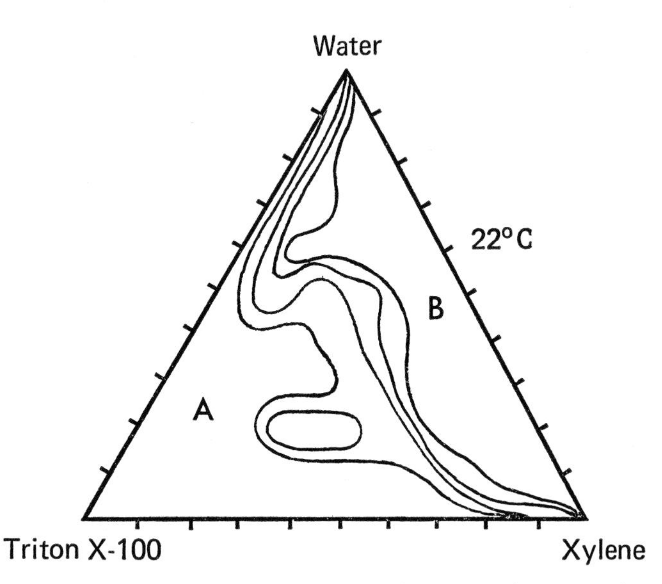

FIG. 10. The phase "diagram" of all combinations of water, Triton X-100 and xylene at 22 °C. The areas outlined range from completely clear (A) to completely opaque (B). (Fox, ref. 9).

Van der Laarse proposed the use of a triangular plotting system, analogous to a phase diagram, to determine the correct counting conditions for ground waters containing tritium. The approach has been adopted by Fox,[9,59] who has produced "phase diagrams" for a variety of detergent + solvent + sample systems. The samples studied so far are sucrose,[88] urine, Fischers medium + horse serum, human plasma and TYG microbiological medium. An example of the plot for Triton X-100 + water + xylene is given in Fig. 10.

The diagrams are constructed by arranging 36 vials in a triangular order. A standard addition of concentrated solute solution (say 0.5 ml of 50 g/l PPO in toluene) is made to each vial and then the volume in each vial is made up to 10 ml by varying the amounts of each of the three components in an incremental manner along the rows defined by the scale along each triangle edge. To each vial is added a 10 μl aliquot of standard tritiated water. The vials are now shaken and counted several times over a 24 h period. The phase diagram is now completed by first visually assessing the degree of opacity of the solution in each vial on an arbitrary scale (say clear = 0 to very cloudy = 4) and second counting each vial to produce a merit value. The merit value is assessed as the percentage counting efficiency times the percentage (by volume) of water present. (This is not the same as the figure of merit, p. 27.)

These two parameters are superimposed in a diagram, as shown in Fig. 11, so that a suitable optimization of sample stability with high merit value can be made. To these criteria may be added a third, namely that it is advisable to count clear solutions because they are true colloidal solutions, even though these may not correspond to the best counting conditions available, i.e. an opalescent solution may have the highest merit value. A final refinement can be to take the optimal vial mixture and make up several samples of this mixture with varying proportions of concentrated solute in toluene solution with the possibility of further increasing the merit value. The clear areas in Fig. 10 are true colloidal solutions whereas opaque areas are emulsions, and the readers attention is drawn to the comments made in other chapters (pp. 27, 28, 31, 58 and 62) relating to possible errors in the counting of inhomogeneous samples.

Rapkin[43] has reviewed gel and emulsion counting and draws some salient conclusions, namely that, for small solution volumes, Cab-O-Sil may be the simplest and most economical agent with toluene whereas larger volumes will require the use of Triton X-100. He also points out that controlled sub-ambient temperature counting is imperative in emulsion colloid samples and that although chemiluminescence is not a large problem in this method it exists nevertheless, and can be controlled by dark adaption and suitable pH control.

Leiberman and Moghissi[83] found that working in red light eliminated chemiluminescence in emulsified samples. They

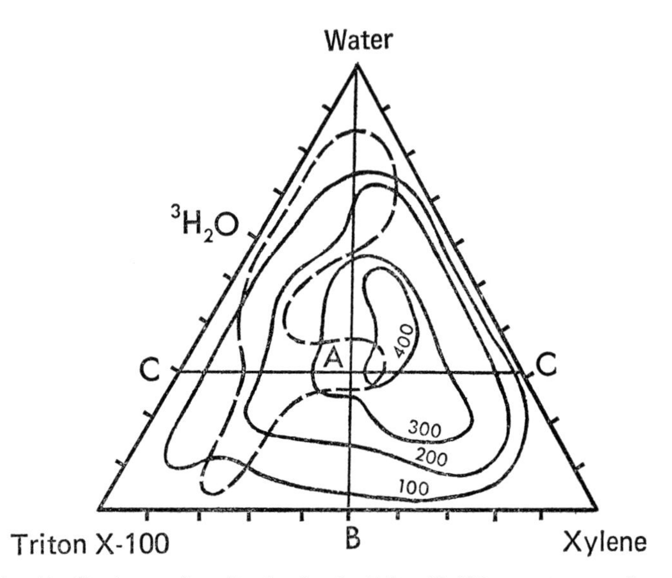

FIG. 11. Contours of merit value for the Triton X-100 + water + xylene system (in increments of 100) superimposed on area of comparative counting stability (dashed line). Point A represents final counting conditions. Ratio of Triton X-100 to xylene in cocktail is as at B and amount of water as at C (Fox, ref. 9).

tested some 40 emulsifiers for use with toluene in the determination of tritiated water. They selected a mixture of 2.75:1 *p*-xylene, Triton N-100 with 7 g/l PPO and 1.5 g/l bis MSB as the best sample preparation. This will take up 10 ml of water into 15 ml of solvent + detergent to give a ^3H counting efficiency of 24% in polythene vials with backgrounds in the range 9–15 counts/min.

3. Paper chromatograms

Separation by paper chromatography often is an important part of the assay of labelled substances. In 1961 Davidson[20] demonstrated that radioassay of "spots" on a paper chromatogram could be carried out by cutting a strip containing the

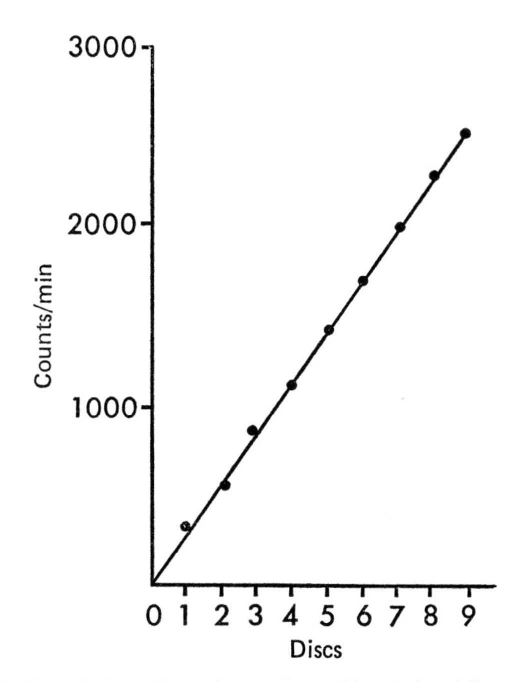

FIG. 12. Cumulative effect of counting tritiated thymidine dried onto glass fibre discs in the same vial (Fox, ref. 9).

"spot" and placing it into a liquid scintillant. This simple method is reproducible and sensitive for ^{14}C but less reliable for ^{3}H labelled materials. The method used today is to arrange the paper strip at the bottom of the vial. This is a more reproducible location, though a less sensitive one, than placing the strip vertically in the vial.

4. Disc-counting

The counting of substances deposited on filter paper is an attractively simple method of sample preparation and has been extended in a number of novel ways.

One way is to dry solutions of tritiated substances on to filter paper discs. These discs are easy to prepare and require only a minimum of toluene-based scintillator solution to wet them enough to count. Up to 25 such discs can be counted in one sample container, which is a useful way of collecting significant counts from a low specific activity sample. The method is reproducible as can be seen from Fig. 12. Recent trends have been towards the use of glass fibre discs, rather than cellulosic filter paper, as glass fibre discs promote higher counting efficiencies. Users of this method must be aware that small labelled molecules will diffuse into the absorbent materials resulting in reductions in efficiencies via self-absorption of β energies. This source of error is less apparent on glass fibre discs than on some filter papers.

Another variant of this technique is to use ion-exchange papers (e.g. DEAE) to take up charged biological species for counting, and thus discriminating against the uncharged species present. Many enzymic assays have been accomplished in this manner, but conditions for salt elution and paper drying can be critical. A nuance of this approach is to contain samples on Millipore filters.[62] This has been advantageous in the determination of biopolymers, sub-cellular particles (e.g. ribosomes) or even algae.

5. Minibags

Gupta[8] has devised a cheap and simple method in which steroids, labelled with ^{3}H or ^{14}C, are deposited onto fluoro-coated glass fibre discs. These are then dried and encapsulated in a plastic bag with 10 to 100 μl of toluene. The bag then is supported in a vertical position in a vial for counting. He also proposes that this is a useful general method for homogeneous samples and advises 50 μl aliquots containing 1 to 10 μl of radioactive sample as an organic or aqueous phase.

6. Polyacrylamide gels[6]

The use of electrophoretic separations of proteins in polyacrylamide gels has provoked interest in the direct counting of ^{14}C, ^{3}H and ^{32}P labelled macromolecules in gel slices. Solubilization by NCS has been popular, but Instagel and Biosolv BBS-3 are said to be preferable solubilizing agents. Combustion may be superior to solubilization.

C. Oxidation techniques[6,26,37,38]

1. Introduction

The foregoing information in this chapter has illustrated the incorporation of a wide sample variety into liquid scintillation solutions. The major problems which can arise from homogeneous and inhomogeneous sample preparation can be summarized as: (i) preparation can be lengthy and expensive; (ii) use of blending solvents and solubilizers can, and generally does, reduce the observed quantum efficiency of the solute by quenching; and (iii) preparation can induce luminescent processes.

One comprehensive solution to these problems is to oxidize the sample to $^{14}CO_2$ or tritiated water. These products are colourless and can be taken into a cocktail in such a way as to be determined in a reproducible manner. The element of time is an important factor in radioassay, and users of automatic instruments are justifiably concerned to make the best use of these expensive machines. Shrewd analysis must include a justification for complex sample preparation methods set against the knowledge that they may create quenching and/or chemiluminescence. Both phenomena cause time delays: quenching by the longer counting times needed to produce acceptable counting statistics and chemiluminescence by further sample handling and/or "dark adaption" to allow luminescence to decay. The result of this analysis may well be that oxidation is the most cost-effective solution when the samples are known to be highly quenching or are not amenable to homogeneous or heterogeneous sample preparation.

In the special case of samples labelled with both ^{3}H and ^{14}C ("dual-labelled"), quenching has the specific difficulty of increasing the overlap of the β spectra (Fig. 13). Here oxidation provides the possibility of determining the isotopes separately by differentially trapping out the oxidation products. The disadvantage of oxidation techniques is usually that of limited sample throughput, but automatic combustion is a possibility which will be considered later. The available oxidation techniques will now be discussed.

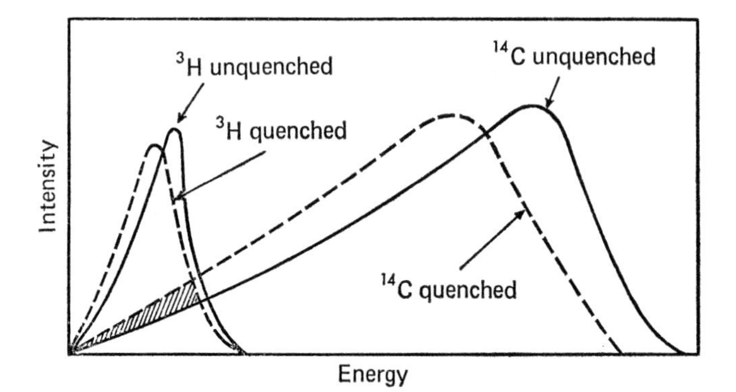

FIG. 13. Diagramatic representation of the increase in spectral overlap (shaded area) caused by quenching (attenuated spectra).

2. Wet chemical oxidation

Samples oxidized, first, by perchloric acid and then by hydrogen peroxide followed by heating at 70 to 80 °C, can be taken into a toluene + PPO phosphor with "Cellusolve" (note POPOP causes yellowing). This method can be used for ^{14}C and ^{3}H in proteins, blood, urine, soft tissues, bone, teeth, faeces and other biological substances. The treatment will also digest Millipore filters and some chromatographic papers. The method is that of Mahin and Lofberg based on the methods pioneered by Jeffay and others. When large volumes of sample need to be handled, additions of NCS or "Instagel" have been advised. The method can be used for isotopes other than ^{14}C and ^{3}H (e.g. ^{45}Ca, ^{55}Fe, ^{32}P and ^{35}S).

3. Sealed tube oxidation

Many methods have been described whereby the sample is sealed into a tube with an oxidation catalyst (e.g. CuO + Mn O_2 + $CuCl_2$ mixture). The tube is heated to 850 °C in a furnace, broken open, and the vapour produced collected in a cold finger (liquid nitrogen for $^{14}CO_2$ and dry-ice + acetone for THO). Gases collected in this way can then be distilled into a suitable trapping agent in a scintillator solution. The method is rapid but sample size is limited and the recovery of combustion products can be as low as 75%.

4. Oxygen train combustion

A standard apparatus is available to convert carbon and hydrogen compounds to CO_2 and H_2O by carrying them in a stream of oxygen over a Ni + CuO catalyst at 930 °C. The recovery is good but the time of analysis is long by comparison and the method is liable to memory effects.

5. Oxygen bomb methods

A modified conventional bomb calorimeter can be used to combust non-volatile material. Usually the bomb is cleared after combustion by evacuation, and the total gases condensed out for trapping and counting.

6. Oxygen flask methods

Combustion of organic materials in a flask containing oxygen was well known prior to the advent of liquid scintillation counting, by virtue of the "Schoniger flask" method. The technique is a simple one in which samples in a platinum basket are ignited in oxygen. The original method of ignition was by electric discharge but a more recent, and safer, method is to ignite by infrared light. In current practice, banks of six, or more, flasks are ignited simultaneously. The flasks are designed with a finger at the bottom which can be immersed in a Dewar flask containing dry-ice + acetone during combustion. The flasks contain about 10 ml of an absorption mixture to trap the gases produced in a flask of about 1.5

litre volume. Reproducibility and recoveries are good and low memories usually are observed. Liquid and solid samples can be combusted in cellophane bags.

7. Trapping agents

All combustion procedures require that the gaseous products be trapped into the scintillator. For THO a PPO + POPOP in toluene solution is adequate with 30% of methanol added as a blender. Carbon dioxide presents a different problem and earlier methods suggested alcoholic potassium hydroxide, Hyamine or ethanolamine as trapping agents. Another approach was to precipitate barium [^{14}C]carbonate for counting. These ways are being replaced by phenylethylamine (scintillator grade) and NCS. (Note: carbon dioxide is taken into phenylethylamine by carbamate formation. It has been reported that the carbamate is unstable and should be counted immediately.)

8. Automatic combustion equipment[6,90]

There is a strong body of opinion amongst experienced users of liquid scintillation methodology that the combustion technique is the most accurate way of sample preparation for all biological samples and highly coloured substances. This is a truism and the prospect of its wider acceptance has prompted commercial enterprise to design combustion accessories which can be combined with existing scintillation spectrometers. This provides an automatic sample combustion, trapping and counting system capable of an acceptable sample throughput.

Recent re-assessments, of this long-standing objective, have produced commercial instruments based on either a catalytic oxidation method or a reliable automatic combustion analogous to the Schoniger technique. Both designs are proving to be more reliable than their predecessors and can handle about 160 samples per day. A foreseeable consequence of these recent developments is that a truly reliable automatic combustion accessory should reduce the complexity of the basic spectrometer in respect of the number of channels and the data handling needed. It would also make the choice of a suitable cocktail from the hundreds described in the literature a simpler one!

D. Low level counting[6,9,24,28,83]

1. Carbon dating

Liquid scintillation gradually is replacing gas flow counting as a method for the estimation of low level ^{14}C in dating studies. This requires the conversion of ^{14}C archaeological samples into a form compatible with the appropriate solutes and solvents. Methods of conversion to $^{14}CO_2$, acetylene,

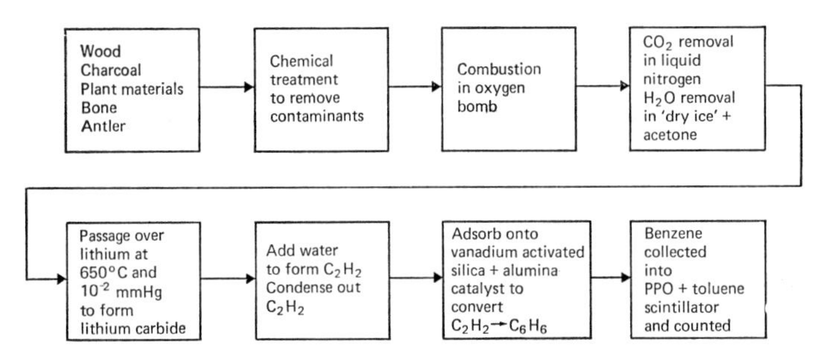

FIG. 14. Plan of possible method to determine ^{14}C content in archaeological samples for dating.

benzene, methyl alcohol, hexane, octene, paraldehyde and toluene all have their protagonists. Probably the most favoured method is one after the scheme outlined in Fig. 14. For precise work the performance of the scintillation method is consistently better than the gas proportional counting technique.

2. Tritium dating

The monitoring of atmospheric tritium is a problem arising from nuclear weapon testing. Tritium concentration also can be used to determine the age of ground waters or the influence of environmental factors on the rates of flow or rain water into underground reservoirs. The use of vintage wines as standards is an attractive bonus to certain research methods.

The parameter used is that of the tritium unit (T.U.) with one T.U. corresponding to the ratio of one tritium atom to 10^{18} hydrogen atoms, i.e. 1 T.U. = $^{3}H \times 10^{-18}/H$. There are three major methods for the determination of low tritium

TABLE VII. Comparison of methods for low level ^{3}H counting

Method	*Counting efficiency (%)*	*Detection limit (T.U.)*
Conversion to benzene	30	6–10
Water incorporation	10–20	20–72
Exchange (toluene)	25	22

activities, one which depends upon its conversion to ^{3}H benzene, another relying on catalysed exchange between tritium in the sample and toluene (or another hydrocarbon), and finally the direct incorporation of water samples by a dioxane + naphthalene + PPO + POPOP cocktail. Examples of the detection limits and counting efficiencies of these methods are in Table VII.

Chapter 5

THE USE OF LIQUID SCINTILLATION COUNTING TO DETERMINE ISOTOPES OTHER THAN CARBON-14 AND TRITIUM

A. Introduction

Some 50 isotopes have been estimated by liquid scintillation or closely related procedures (Fig. 15). The advantages of using liquid scintillation in preference to other methods for isotopes other than ^{14}C and ^{3}H may accrue from the following: (i) the speed and reproducibility of sample preparation, (ii) the counting geometry will maximize counting efficiencies when dealing with low activity samples or low energy emitters, (iii) the convenience of use with small volumes of solution, (iv) the convenience in sample handling and data analysis provided by the automatic instruments, (v) the possibility of adaptation to flow cell use, (vi) the sensitivity to concentration which enables the technique to be used for trace materials, and (vii) the suitability of the method for fundamental studies of nuclear processes and absolute counting, Chapter 1 contains mention of the energetic processes, other than β^- emission, which activate the scintillation process and thus widen the scope of the method.[60]

B. Fundamental studies

As most of the earlier literature relating to inorganic isotope determination in scintillator solution is concerned with nuclear studies a brief survey of this work follows.

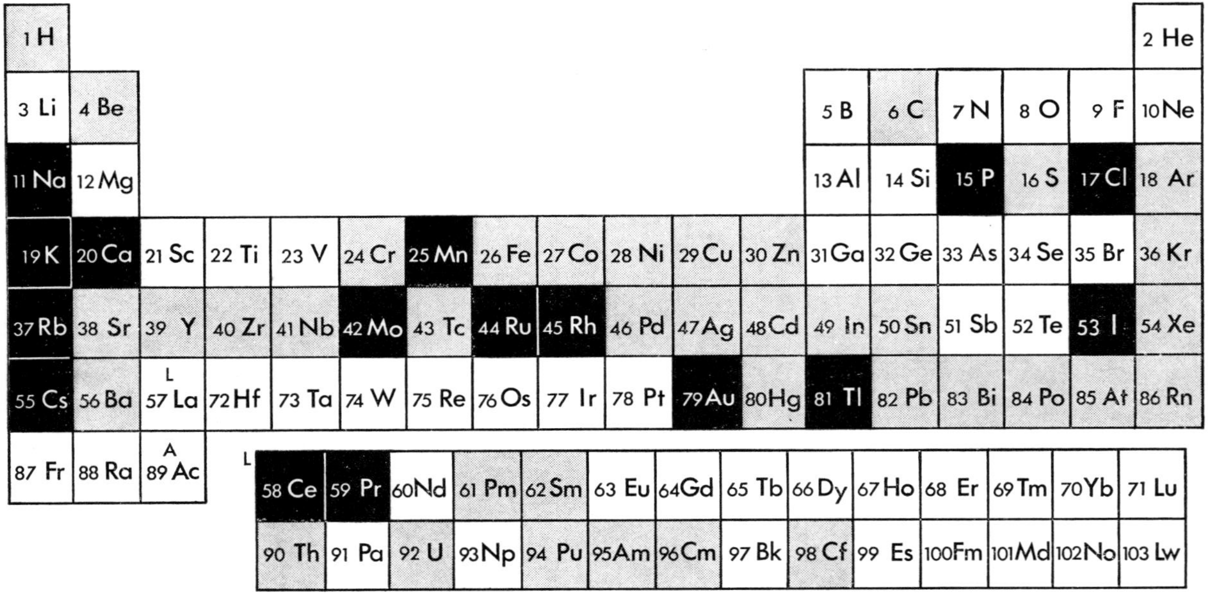

1 H																	2 He
3 Li	4 Be											5 B	6 C	7 N	8 O	9 F	10 Ne
11 Na	12 Mg											13 Al	14 Si	15 P	16 S	17 Cl	18 Ar
19 K	20 Ca	21 Sc	22 Ti	23 V	24 Cr	25 Mn	26 Fe	27 Co	28 Ni	29 Cu	30 Zn	31 Ga	32 Ge	33 As	34 Se	35 Br	36 Kr
37 Rb	38 Sr	39 Y	40 Zr	41 Nb	42 Mo	43 Tc	44 Ru	45 Rh	46 Pd	47 Ag	48 Cd	49 In	50 Sn	51 Sb	52 Te	53 I	54 Xe
55 Cs	56 Ba	L 57 La	72 Hf	73 Ta	74 W	75 Re	76 Os	77 Ir	78 Pt	79 Au	80 Hg	81 Tl	82 Pb	83 Bi	84 Po	85 At	86 Rn
87 Fr	88 Ra	A 89 Ac															

L	58 Ce	59 Pr	60 Nd	61 Pm	62 Sm	63 Eu	64 Gd	65 Tb	66 Dy	67 Ho	68 Er	69 Tm	70 Yb	71 Lu
	90 Th	91 Pa	92 U	93 Np	94 Pu	95 Am	96 Cm	97 Bk	98 Cf	99 Es	100 Fm	101 Md	102 No	103 Lw

FIG. 15. Periodic table illustrating those isotopes determined by liquid scintillation counting (grey) and those determined by Čerenkov counting (black).

The technique has been widely used for the absolute counting of α emitters. A prominent worker in this field has been Horrocks who, with his co-workers,[18,77] has studied ^{217}At, ^{242}Cm, ^{232}Th and isotopes of uranium and plutonium. He,[73] and other workers, have examined many β^- emitters, for example ^{137}Cs/^{137m}Ba, ^{131}I, ^{241}Pu, ^{35}S and ^{90}Y. Absolute counting efficiencies close to 100% can be recorded.

Various half-life studies are known and Horrocks[75,76] studied pulse-height–energy relationships for electrons and photo-electrons in liquid scintillation media for about 16 isotopes.

All these investigations necessitate the incorporation of inorganic isotopes into a solute + solvent solution and often form the basis for routine sample preparation. This aspect is covered in the subsequent text and summarized by appropriate tables.

C. Determination of solid samples

The methods used follow the same general lines as those for the low energy β emitters ^{14}C and ^{3}H described in Chapter 4.

1. Cab-O-Sil

This has been a popular agent for dispersing finely divided solids in a phosphor. Unicellular algae, containing ^{45}Ca and ^{89}Sr, has been counted in this way and the same isotopes have been recovered as solids from urine, plasma and bone (as ash) to be estimated in the same way.

The counting of iron isotopes (^{55}Fe, ^{59}Fe) has been a problem which can be overcome by counting as precipitates dispersed in a toluene based scintillator by Cab-O-Sil. Ferriphosphate and ferric benzene phosphinate precipitates can be used successfully in this context.

2. Polymer gels

Methyl methacrylate and polyolefinic gels (Poly-Gel-B) have been used to suspend inorganic solids in PPO or butyl-PBD/POPOP/toluene phosphors. Suspension is achieved by warming the phosphor, polymer and sample. Some inorganic solids measured in this way include the sulphates of ^{45}Ca and ^{90}Sr, the perchlorate of ^{137}Cs, sodium [^{131}I]iodide and the complexes of iron mentioned previously.

3. Filter paper

Deposition of precipitates on glass fibre filter paper followed by mulching of the paper and precipitate in a toluene scintillator is a method for ^{35}S as barium sulphate and ^{32}P and ^{45}Ca as calcium phosphate. Studies on ashed tissue and bone (containing ^{45}Ca) show that better results are obtained by the use of cellulosic papers, in these cases, as it is less prone to shed lint.

Another method in the literature uses an anionic exchange paper dispersion to estimate ^{99}Tc and this approach may be worthy of further study. Note: when all else fails results can be obtained by placing the labelled solid directly in the scintillator, and this may be preferable to embarking on a complex chemical separation to recover the isotope in a "countable" form. Obviously this crude method stands more chance of success with higher-energy emitters but then Čerenkov counting may be an alternative (p. 89).

D. Determination of liquid samples

Again principal methods follow those for ^{14}C and ^{3}H except that examples of labelled compounds soluble in scintillator solvents are almost unknown. The only recorded instance is the uptake of silicon [^{36}Cl] chloride into toluene with a PBD solute. This is used in a dating technique for sodium chloride in ground waters.

1. Aqueous solutions of inorganic salts

Several recipes for ^{45}Ca in samples of biological origin (serum, urine, bone, blood and milk) involve various treatments to recover a precipitate of calcium oxalate. This then is dissolved in an aqueous mineral acid and blended with ethanol into a toluene scintillator. Similar methods have been evolved for the iron isotopes recovered as chloride or perchlorate. Ascorbic acid can be added to stabilize the iron in its ferrous state. Where larger volumes of solution or a higher concentration of salt are present cocktails of the dioxane + glycol type can be used.

2. Use of a complexing or solubilizing agent

These methods depend upon the use of an organic material to produce a complex, or an ion-association complex, of a metal isotope which then has sufficient scintillator solvent solubility. Clearly a colourless end-product is advisable and phosphate complexing agents have been preferred for this reason. An alternative may be to use a complexone such as ethylenediamine tetra-acetic acid (EDTA). Table VIII contains a summary of these methods.

3. Use of the organic salt of the metal isotope

The salts of metals with organic acids often have appropriate solubilities in the alkylbenzenes to facilitate liquid scintillation counting (Table IX).

Some authors have found it an advantage to add further emulsifying or solubilizing agents to the counting mixtures. These are summarized in Tables X, XI with other schemes where a solvent extraction step has proved useful.

E. Determination of gaseous samples

The extension of the method to the estimation of radioactive gases can be accomplished either by trapping the gas or by relying on the gas having a reproducible solubility in a suitable solvent (usually toluene). Trapping techniques have been used for $^{35}SO_2$, the gas is contained in $Na_2Hg_2Cl_4$ or H_2O_2 + H_2SO_4 solutions and counted in a PPO + DMPOPOP + naphthalene + dioxane cocktail. The combustion methods can be used to produce $^{35}SO_2$ when ethanolamine is recommended to be the best trapping agent, although a Triton X-100 + toluene mixture has been very

TABLE VIII. Use of complexing agents

Isotopes	*Cocktail*	*Experimental details*	*Comments*
^{59}Fe	PPO + POPOP (Hayes and Gould)	Fe orthophenanthroline complex, 50% efficiency	Iron metabolism studies
^{55}Fe	DMPOPOP + PPO + naphthalene + toluene	Fe di(2-ethylhexylphosphoric acid) (HDEHP) complex	Blood
^{63}Ni	NE 240	(Ni py_4)$(CNS)_2$ complex	Effluent from AGR reactors
^{249}Pu, ^{241}Am	PPO + POPOP + naphthalene + dioxane	Tri-*n*-octylphosphine oxide (TOPO) complex	Comparison with other methods
^{249}Pu	*p*-Terphenyl + POPOP + toluene	Acid extraction from ashed samples by HDEHP	Biological samples
^{241}Pu, ^{63}Ni, ^{151}Sm, ^{35}S, ^{106}Ru + ^{106}Rh	*p*-Terphenyl + POPOP + xylene	Sm, Pu as HDEHP in HCl, ^{63}Ni as di-octyl phosphate complex, S as Na_2SO_4, Ru + Rh *p*-toluidine complex	Absolute disinteg. studies
^{239}Pu, ^{242}Cm, ^{232}Th, ^{233}U, ^{217}At, ^{252}Cf, ^{90}Y	DMPOPOP + PPO + toluene	HDEHP extraction in xylene. Quenching studied	α Resolution studies, etc.
^{239}Pu, ^{106}Ru, ^{144}Ce, ^{95}Zr, U	DMPOPOP + PPO + naphthalene + dioxane	TOPO-carrier free, comments on stability	

^{144}Ce, ^{60}Co, ^{22}Na	POPOP + PPO + naphthalene + dioxane + xylol + ethanol	HDEHP complexes	Counting loss studies
^{147}Pm	DMPOPOP + PPO	HDEHP complex	Urine assay
^{147}Pm	POPOP + PPO + naphthalene + dioxane	HDEHP in 0.1 m HNO_3	Fission products
^{241}Am, ^{244}Cm ^{152}Eu	PPO + POPOP	HDEHP complex form aqueous oxalate and sulphate solution at pH 3.6 and constant ionic strength	Ion-exchange effiuent
^{232}Th, ^{63}Ni, ^{109}Cd, ^{133}Ba, ^{241}Am, ^{113}Sn, ^{113m}In, ^{203}Hg	*p*-Terphenyl + POPOP + xylene DMPOPOP + PPO + toluene	HDEHP complexes	Half-life and other fundamental studies
^{232}Th	*p*-Terphenyl + POPOP + toluene	DOP complex	

Specific references can be found in Chapter 9 of ref. 9 in the Bibliography.

TABLE IX. Use of metal organic salts

Isotopes	*Cocktail*	*Experimental details*	*Comments*
$^{137}Cs + ^{137m}Ba$	DMPOPOP + PPO + toluene	Salt of 2-ethylhexanoic acid (octoate)	
^{87}Rb	*p*-Terphenyl + POPOP + toluene	Octoate	Half-life studies
^{45}Ca	PPO + POPOP + toluene + ethanol	Octoate from oxalate in HCl	Biological fluids
^{147}Sm	*p*-Terphenyl + POPOP + toluene	Octoate	Half-life studies
^{63}Ni	PPO + POPOP + toluene + ethanol	*n*-Caproate	
$^{90}Sr + ^{90}Y$	PPO + POPOP + toluene	Octoate	Assays

Specific references can be found in Chapter 9 of ref. 9 in the Bibliography.

TABLE X. Complexing or extraction with emulsifier added

Isotopes	*Cocktail*	*Experimental details*	*Comments*
^{32}P	PPO + POPOP + toluene	Phosphomolybdic acid, extracted *n*-butanol + ethyl ether + ethanol Add Hyamine 10X	Food analysis
^{45}Ca		Add Triton X100	Urine and serum
^{45}Ca, ^{55}Fe, ^{87}Rb, ^{147}Sm, ^{90}Sr, ^{89}Sr, ^{36}Cl	PPO + bis MSB + xylene	Ca, Fe as fluoride, Rb, Cl, Sr, Y as chloride, Pm coppt. as oxalate with praesodymium, then as EDTA complex All with Triton N-101 added	Low-level counting
^{55}Fe, ^{59}Fe	PPO + POPOP + toluene	Ferrous citrate + Hyamine 10X	Plasma samples
^{51}Cr, ^{54}Mn, ^{55}Fe, ^{65}Zn, ^{88}Y, ^{125}I	PBD or PPO + POPOP	Cr as chloride or chromate, Mn as nitrate, Fe, Zn, Y as chloride, I as potassium iodide, Triton X100 added	Electron capture nuclides
^{60}Co	BBOT + toluene	CO^{II} thiocyanato complex with Triton X100	Low-level counting of environmental waters
^{51}Cr		Cr as EDTA complex. Triton X100 added	Simultaneous measurement with ^{14}C

Specific references can be found in Chapter 9 of ref. 9 in the Bibliography.

successful. The commercial instruments have not been recommended at this time for ^{35}S oxidation.

Sulphur labelled hydrogen sulphide can be trapped in thiosulphate solutions prior to counting in a Hyamine hydroxide cocktail but, since this gas has a good solubility in toluene, a simpler method is to inject it into a serum-capped vial filled with a toluene based scintillator.

The noble gases dissolve in toluene and ^{37}Ar (Iorgulescu[78]), ^{85}Kr, ^{222}Rn and ^{131m}Xe (Horrocks and Studier[74]) have been estimated by using a pre-evacuated vial or ampoule into which the gas and scintillant are injected or pre-frozen. PPO and POPOP are the preferred solutes.

F. General comments on the counting of metal isotopes

1. Solutes

PPO and POPOP are often compatible with inorganic samples and can tolerate quite stringent conditions of acid and alkali. TP can be a cheap alternative, particularly with strong β emitters counted at ambient temperatures.

2. Solvents

Toluene, xylene and dioxane seem adequate in most cases. Recently, Gomez has suggested the use of benzo- and acetonitriles as general scintillator solvents and reported their use with ferric, mercuric, antimony and copper chlorides and zinc acetate.

3. Vials

The use of siliconized vials may be necessary to reduce the adsorption of metallic isotopes onto glass.

4. Quenching

The fluorescence produced by the solute can be reduced by the presence of acids, alkalis, inorganic salts, solubilizers and complexing agents. It also can be reduced by coloured

TABLE XI. Solvent extraction procedures

Isotopes	*Cocktail*	*Experimental details*	*Comments*
^{45}Ca	PPO + toluene	Ca perchlorate extracted by tributyl phosphate (TBP)	Biological sources
^{45}Ca	PPO + αNPO + toluene *p*-Terphenyl + αNPO + toluene	Ca chloride extracted by dibutyl phosphate (DBP). Quenching studies	Agricultural materials, plants and soils, etc.
^{90}Sr + ^{90}Y		DBP extraction	
^{241}Pu	*p*-Terphenyl + POPOP + xylene	DBP extraction from 1 M HCl	
^{90}Y, Pu		TBP extraction	Agricultural materials
^{241}Am, ^{233}U	DMPOPOP + PPO + naphthalene + dioxane or BBOT + toluene	Tri-isoctylamine + 30% DMF extraction	
Th, U	*p*-Terphenyl + αNPO + toluene	TBP extraction	

Specific references can be found in Chapter 9 of ref. 9 in the Bibliography.

solutions. These problems are most acute when estimating low energy β^- emitters (e.g. ^{45}Ca) and electron capture nuclides (e.g. ^{55}Fe, ^{125}I). Quenching can be allowed for by the standard methods of quench correction developed for ^{14}C and ^{3}H, but highly coloured solutions should be avoided.

5. Chemiluminescence

The presence of inorganic species has been suspected as a source of chemiluminescent reactions in certain cocktails. It

TABLE XII. Multiple inorganic isotope counting[6,9,67]

Isotopes	*Comment*
$^{55}Fe + ^{59}Fe$	Several methods available for blood samples
$^{125}I + ^{129}I + ^{131}I$	Important in medical use of radioisotopes (e.g. thyroid uptake) PPO + POPOP in toluene used usually
$^{95}Zr + ^{95}Nb$	Dibutyl phosphate extraction into TP + POPOP in xylene for absolute standardization
$^{32}P + ^{45}Ca + ^{89}Sr$	Bone metabolism studies. Bray cocktail for aqueous samples (Ca, Sr as perchlorates)
$^{45}Ca + ^{90}Sr$	Faecal and urine ashed samples converted to chlorides, butyl-PBD + toluene + ethanol cocktail
$^{239}Pu + ^{241}Pu$	Urine samples. Ferriphosphate complex counted as aqueous samples in a Cab-O-Sil gel
$^{45}Ca + ^{89}Sr$	Plant and soil samples. Counted as di-*n*-propyl phosphate extract with PPO + α-NPO or TP + α-NPO in toluene scintillator

TABLE XIII. Comparison of methods for the detection of inorganic isotopes

Isotope	*Liquid scintillation*		*Čerenkov*		*Well-crystal scintillation (3 inch diameter)*		*End-window G-M*		*Gas flow (anti-coincidence)*	
	B†	E†	B	E	B	E	B	E	B	E
^{32}P (β 1.7 MeV)	20	90	12	40	40	10	15	40	0.5	45
^{125}I (γ 0.027 MeV)	20	90	—	—	35	65	15	2	0.5	2
^{51}Cr (γ 0.32 MeV) (X-ray 0.005 MeV)	20	10	—	—	30	55	15	1	0.5	1
^{210}Po (α 5.29 MeV)	20	95	—	—	—	—	15	10	5	30

†B, Background (counts/min); E, Efficiency (%).

has been observed in toluene + ethanol mixtures (with Zn and Ba) and in dioxane (with Mg).

6. Multiple isotope counting

A variety of inorganic isotopes have been simultaneously determined. The analysis requires careful setting of instrument windows and a knowledge of the modes of decay of the isotopes to be determined. Some examples are in Table XII. See also p. 69.

7. Comparative studies

Some authors[9] have made detailed investigations of the liquid scintillation method in comparison with other techniques for the counting of inorganic isotopes. These generally have been favourable and some are summarized in Table XIII.

Chapter 6

THE PHENOMENON OF QUENCHING AND ITS CORRECTION

A. Introduction

Previous chapters have discussed the addition of various materials to scintillator solutions to prepare samples suitable for liquid scintillation counting. All these agencies will alter the performance of the solute to some degree, and so may the sample itself. Those materials which have a minor effect are listed as "diluters" in Table II. The effect of diluters can be restored, in part, by increasing the concentration of the primary solute. Other samples, secondary solvents, solubilizers, emulsifiers etc., may have a more serious effect on the light output of the solute. These are known as quenchers.

B. Types of quenching[6]

Three types of quenching can be defined and any one substance may quench by more than one type.

1. Impurity quenching

This is more often described as chemical quenching but this is really a misnomer as its source is a physical effect rather than a chemical one. It arises when the fluorescence quantum yield is decreased by the presence of non-fluorescent molecules which compete with the solute molecules for the excitation energy of the solvent molecules.

These non-fluorescent molecules dissipate the energy by radiationless transfer (i.e. convert translational and vibrational energy to heat). They also can combine with solvent or solute molecules to form encounter complexes which consist of either unexcited molecules (static quenching) or excited molecules which undergo deactivation reactions (dynamic quenching).

The formation of these complexes is a diffusion controlled process (as recently shown by Birks[6,9]) and the resultant magnitude of quenching depends upon (i) the solvent excitation lifetime, (ii) the solute fluorescent lifetime and (iii) the chemical nature and molar concentration of the quencher. Note that the production of these complexes can be discouraged by the structure of the solute: hence the trends in the design of new solutes previously commented on. Impurity quenchers can be roughly divided into those with a mild effect (R—Cl, RCOOH, R—NH_2, R—CH = CH—R, R—S—R) and those with a more severe effect (R—Br, R—SH, R—OCOCO—R, R—CO—R, R—COX, R—NH—R, R—CHO, R_2N—R, R—I, R—NO_2, O_2); R = aliphatic group, X = halogen.

Colour quenching

Coloured samples can arise from either chemical or photochemical reactions in the scintillator solution promoted by the labelled material, or the incorporation of inherently coloured materials (e.g. urine). The effect is to diminish the mean free path of fluorescence photons with a resultant impairment of light collection efficiency at the photomultipliers.

Photon quenching

The incorporation of intractable substances into a cocktail may result in a heterogeneous counting mixture. In these cases the maximum interaction between the β energy and the solvent + solute is not achieved and this can be described as a type of self-absorption effect.

C. Optimization of quenching

It is often impossible to avoid quenching and some simple

rules can be presented whereby its effects can be minimized. These are (i) by keeping the concentration of quencher low; (ii) by increasing the concentration of primary solute; (iii) by cooling to reduce the diffusion coefficients of quenching molecules and hence the probability of a quench collision, and by employing a solute of short fluorescence lifetime, again to minimize the probability of a diffusional collision. The fluorescence lifetimes of some common solutes are PPO, 1·6 ns; PBD, 1·2 ns, butyl-PBD, 1·2 ns; POPOP, DMPOPOP, 1·5 ns; BBOT, PBBO, < 1 ns.

D. Quench correction methods

Before reading the ensuing discussion of quench correction the reader is advised to bear in mind the comments in Chapter 4 relating to combustion techniques. The advisability of using complex instrumental and mathematical procedures to correct for quenching is questionable for highly coloured and severely impurity quenched samples. Combustion is a more efficient and quicker method for such samples and its use can save expensive computer time. The most common methods of correcting for quenching will now be described.

1. Internal standardization

In this, the earliest method of correction, standard additions ("spikes") of the isotope, in an unquenched form, are added to the sample which is then recounted. Comparison of the count rate before, and after, the spike is added enables a correction for quenching to be made.

The method has the disadvantages of requiring a double count and being dependent on the accurate dispensation of the spike. It is not suitable when two isotopes are being counted simultaneously (dual-label counting). It normally is a destructive method of testing, i.e. the sample count rate cannot be redetermined, but vials are marketed with a special compartment to contain a standard. Here the standard must be a γ emitting isotope and perhaps is more properly described as an external standard. Despite the disadvantages quoted internal standardization is the only *accurate* method for heterogeneous samples and it is the most reliable one for highly quenched samples.

2. Channels ratio method (pulse-height shift method)[31,52]

In 1960 Baillie[49] noticed that the addition of a quenching agent caused the observed pulse height of the β spectrum to shift to a position of lower energy (Fig. 16). The concurrent development of commercial instruments designed to accommodate tritium and ^{14}C spectra in two separate counting

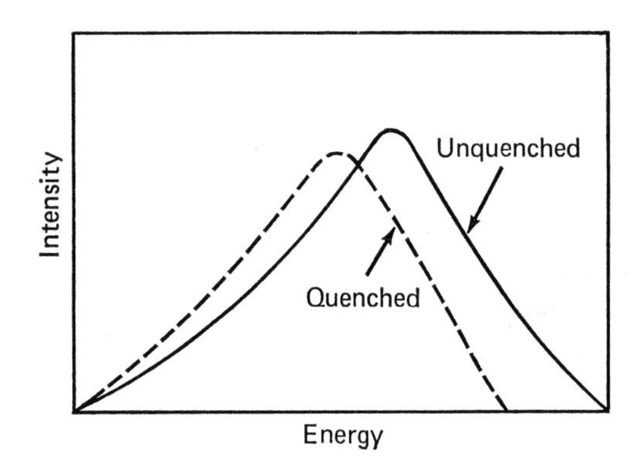

FIG. 16. Spectral shift in β spectrum caused by quenching.

channels was adapted to monitor this shift by selecting two adjacent channels. This was attained by arranging discriminator levels as shown in Figure 17.

The spectral shift is concentration dependent and a calibration curve can be established by counting a series of samples containing a known constant amount of radioactivity but varying concentration of quencher. The modern instrument monitors both channels simultaneously thus reducing statistical errors. The calibration curve is constructed of the ratio of the counts in the two channels ("channels ratio") versus the observed efficiency of the known standard in each sample (Fig. 18).

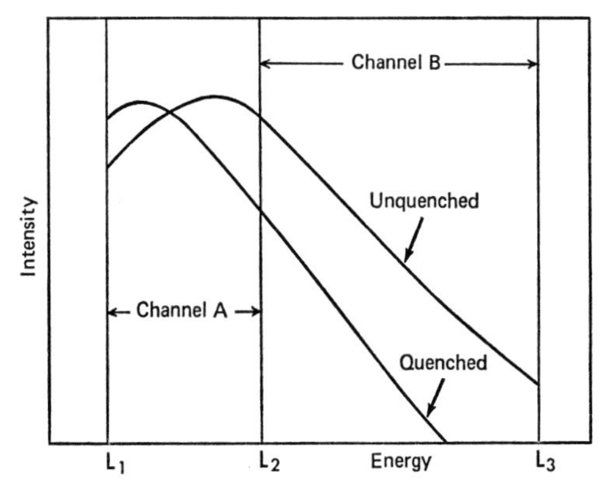

FIG. 17. Quenched and unquenched attenuated β spectra located in two adjacent counting channels (A and B).

The exact meaning of channels ratio is subject to some variation in instrument design but a popular method of channels comparison is to use for ^{14}C the ratio

$$\frac{\text{Counts in channel A } (L_1-L_2)}{\text{Counts in channels A+B } (L_1-L_3)}$$

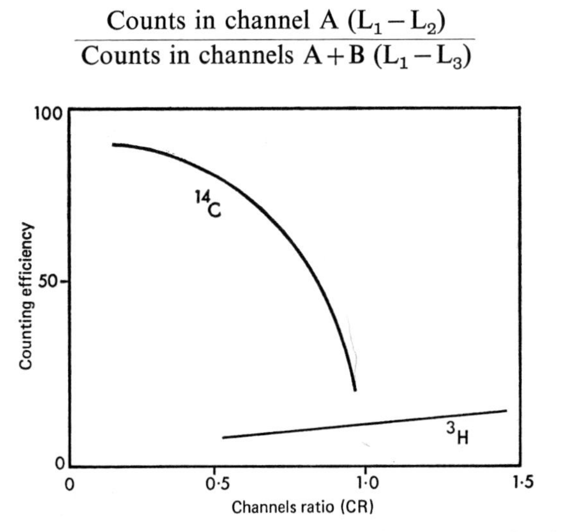

FIG. 18. Quenching curves for ^{14}C (CR = A/A + B) and ^{3}H (CR = B/A).

and for ^3H the ratio

$$\frac{\text{Counts in channel B } (L_2 - L_3)}{\text{Counts in channel A } (L_1 - L_2)}$$

where the letters refer to Fig. 17. These ratios give quench correction curves like those in Fig. 18.

In theory the construction of one calibration curve should suffice for all materials quenching the same isotope, provided that the calibration is for the same scintillator solution and constant instrumental settings. In practice a single curve is obtained only if the mechanism of quenching is solely that of impurity quenching, i.e. that colour and photon quenching are absent. This is a special circumstance and most users calibrate for each particular sample preparation.

The method relies on one measurement, once the calibration curve is known, and it is a non-destructive procedure. Channels ratio can be used for dual-labelled samples and is independent of sample volume. It is not particularly accurate for highly quenched or low activity samples but is suitable for heterogeneous samples.

3. Automatic external standard (AES) or "gross counts" method

An external standard to calibrate for quench correction was first used in 1962.[71] The experimental arrangement relies on the automatic positioning of a γ source close to the vial. The scintillator is activated by Compton and recoil electrons from the γ isotope to provide a standardization. The disadvantages of this early arrangement were that small positional changes in the γ source location, and variations in sample volume and vial wall thickness could cause serious counting errors. In addition the presence of any solid material in the vial or of any surface roughening of the vial outer surface will cause an increase in pulse height per disintegration. This arises from a reduction in totally internally reflected light flashes and causes an increase in counting efficiency. This phenomenon, identified by Gordon and Curtis,[65] is known as the "anti-quench" shift and means that the AES method cannot be used for heterogeneous samples.

The errors arising from the gross counts method can be reduced in two ways. The first method is by the provision of two separate counting windows for the external standard and comparing the counts collected in these two channels to produce quench correction curves. This technique is called "automatic external standard channels ratio" (AESCR). The second method introduces two γ calibration sources, one of high energy (e.g. ^{226}Ra) with one of lower energy (e.g. ^{241}Am) positioned above it to become effective at larger volumes. The outcome of this arrangement is that of a self-compensating system whereby an increase in count rate occurring from an increase in volume absorbing the high energy γ rays is monitored by a concurrent larger absorption of the low energy γ's in the same incremental volume change.

The major advantage of these improved external standardization methods over that of channels ratio is that only short counting times are required for calibration. The channels ratio approach has the inherent snag that it relies upon the counts in one channel being sequentially reduced by the quencher with a consequent increase in counting time being essential to preserve good counting statistics. It is convenient to mention at this point that some authors have introduced the additional definition of channels ratio as "sample channels ratio" to prevent possible confusion with the AESCR method. It also should be mentioned that some instruments use a ^{133}Ba standard which is monitored in the same counting windows as the sample. Those instruments providing the separate standardization channels are sometimes described as "3+2" channel instruments.

The limitations to the general application of AESCR are that it is not suited to the determination of highly quenched samples or experiments where a wide variety of cocktails and counting efficiencies are encountered. It is satisfactory for both single and double labelled counting where the variation in cocktail is small and the samples are not severely quenched.

The AESCR approach is more reproducible than the gross counts method for heterogeneous samples but informed opinion is that is is not truly suitable for calibration in these

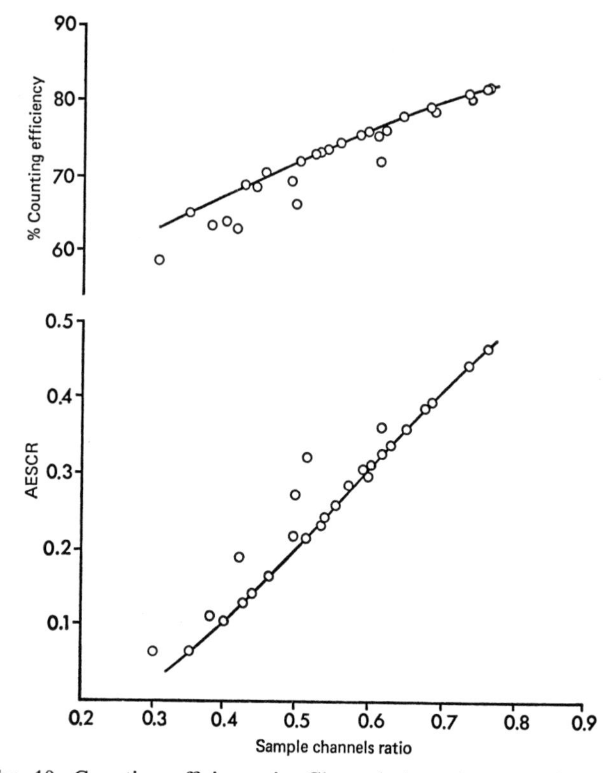

FIG. 19. Counting efficiency in Channel A and automatic external standard channels ratio C/A as a function of sulphur-35 samples channels ratio B/A, where Channel A is a wide (0.5 to 9·9 V) window, Channel B is 3.0 to 9.9 V at the same gain, and Channel C is 9.9 V to ∞ (Bush, ref. 55).

circumstances. This emphasizes the point made in Chapter 4 that it is important to be aware of the exact nature of samples prepared by solubilization and emulsification. A *qualitative* check on sample consistency has been devised by Bush[55] in the "double-ratio" technique. Its basis is that radioactive material which is not truly in solution will give rise to a "degraded" β spectrum due to absorption and back scattering. This will cause the sample channels ratio to be shifted as for a quenching effect (photon quenching). This means that it will not then give the same efficiencies as those measured for the same sample by an external standard method. A plot of sample channels ratio versus AESCR can be constructed for a set of quenched standards. Unknown samples will fall on this curve if true solutions are being estimated whereas deviations will indicate sample inhomogeneity (Fig. 19).

4. Other quench correction methods

One of the earliest methods was the extrapolation routine of Peng (1964).[6] This uses a logarithmic plot of specific activity against quencher concentration to yield a "true" count at zero impurity concentration. It is limited to counting in the integral mode (i.e. with the upper discriminator level set at infinity) instead of the more normal differential mode when the energy spectrum is attenuated between two fixed discriminator settings. The extrapolation approach is not suitable for high degrees of quenching.

Beckman have developed an automatic quench compensation (AQC) which uses electronic servo gain control to restore the shape of a quenched spectrum. The method is reliable and when a single isotope is being counted the instrumental settings can be arranged to over-restore the distorted spectrum. This means that a constant counting efficiency can be maintained over a wide range of impurity concentration.

The major methods of correction are compared in Table XIV.

E. Devising quench correction curves

1. Standards[72]

The most widely used standards are hexadecanes labelled with ^{14}C or ^{3}H. They are preferred to the ^{14}C and ^{3}H labelled

TABLE XIV. Comparison of common quench correction methods[6, 91]

Advantages	*Disadvantages*
	Internal standardization
Simple and reliable. No calibration curve needed. Shorter counting times for low activity sample calibration. Efficiency determined independent of background. Good for highly quenched and heterogeneous samples	Two counts essential (time-consuming). Possible errors due to standard dispensing. Risk of changing efficiency when standard added (i.e. standard will not normally be chemically identical to the sample). Compensation for small volume changes necessary. Cannot recover the sample after standardization. Not suitable for dual-labelled experiments
	Sample channels ratio
Only one count needed. Easy and does not alter the sample. Independent of sample inhomogeneity. Corrects for both colour and impurity quench. Independent of sample volume. Accurate for samples which are not highly quenched and which have reasonable specific activities	Requires calibration curve. Does not get maximum counting efficiency for all samples. Counting time increases with quenching. Not good for highly coloured samples
	External standardization (gross counts and AESCR)
Rapid one count method. Short counting time for low activity samples. Efficiency generally independent of background. Convenient for dual-labelled samples provided care is taken	Gross count method deficient for highly quenched samples and NOT suitable for heterogeneous samples. Sensitive to γ source positioning and to sample volume. AESCR more accurate but still volume and sample dependent. Requires two counts and calibration curves needed

toluene standards, which are available, in that hexadecane has a low vapour pressure at room temperature, simplifying accurate aliquot dispensation. The chemical purity of standards is better than 99% and the accuracy of standardization is within 1·5% for hexadecane and 2% for toluene. Other standards available are tritiated water and ^{36}Cl chlorobenzene.

Manufacturers provide quenched standards usually of hexadecane (^{14}C or ^{3}H) dissolved in toluene and quenched by carbon tetrachloride or chloroform. The solutes are PPO and POPOP. They are provided sealed in argon flushed vials. The provision of aliquots of standards is critical, and the measurement of small quantities of high specific activity liquids calls for care and precision. The preferred method is chemical syringe injection into a serum capped vial.

2. Shape of calibration curves[6,7,82]

Over a limited range of quenching the shape of efficiency versus channels ratio may be a good approximation to a linear curve. The number of points used to define the curve must be adequate and the line can be refined by a least squares analysis.

Similarly, when the curve is not linear, construction must be via a reasonable number of points (up to 80 have been suggested!). Curve-fitting is accomplished by a mathematical curve fitting procedure. Generally a quadratic equation is a good mathematical approximation to the shape of a quench curve, and again least squares analysis refines the "best fit." Some authors have used higher order equations to generate curves but usually this is unnecessary and can result in untenable analyses (Fig. 20).

Linear calibration plots can be obtained by plotting the logarithm of the channels ratio versus counting efficiencies This potentially useful method (Gibson and Gale, 1967) is said to suffer from an insensitivity to quencher concentration.[60]

The growth in use of automatic spectrometers has created a need for the fast handling of data. The best way of improving data throughout is for the individual user to decide. Most laboratories will find that the use of a desk-top calculator or a small programmable calculator will fulfil their requirements, and various electrical interfacing facilities between spectrometer and calculator are provided by the major manufacturers. Certain larger installations can investigate the desirability of using "off-line" or "on-line" computer links. The Intertechnique "Multi-Mat" provides a small "on-line" computer capable of simultaneously processing data from four spectrometers sited within 300 feet.

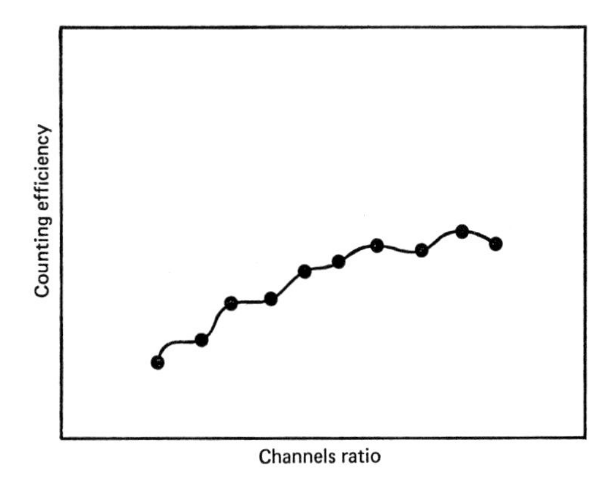

FIG. 20. Quench correction curve generated by computed polynomial "best-fit".

It performs quench correction and standardization procedures from computer stored curves and also computes net counting errors.

Many programmes have been published,[58] both for larger computers and for modern desk-top calculators. Spratt[9] has presented an excellent analysis of the factors and pitfalls in the acquisition and handling of data from liquid scintillation equipment. He strongly emphasises that the ultimate in sophistication can be arrived at by elegant mathematical treatment of results, but this does not improve errors from sample preparation, vial inconsistencies and other experimental factors. Neither do mathematical methods correct for instrument faults.

Glass[7,8,63] has produced a programme which uses a double ratio method. This corrects for quenching by both sample channels ratio and AESCR to allow checks on the most appropriate of these methods for each individual sample. It is a three channel method and contains built-in checks for discrepancies arising from the sample and the counter.

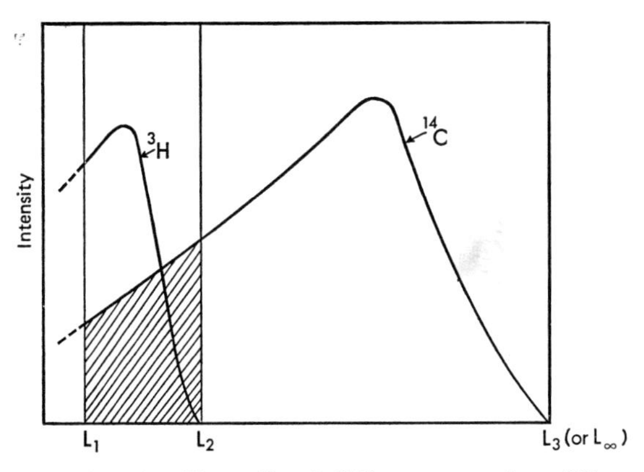

FIG. 21. Overlap ("spill-over") of ^{14}C spectrum into ^{3}H counting window ($L_1 - L_2$).

F. Dual-labelled samples[6]

Modern instruments provide the opportunity to measure two isotopes simultaneously in the same sample. The criterion for this is that a reasonable resolution of their β spectra must be possible. Fortuitously this is possible for the pairs of isotopes which usefully can form part of biomedical applications of radionuclides (^{3}H, ^{14}C; ^{3}H, ^{35}S; ^{3}H, ^{32}P; ^{14}C, ^{32}P). Other less common pairings are included in Table XII.

For dual counting, two channel methods are preferred and the major problem encountered is that the lower energy isotope cannot be estimated without part of the β energies from the isotope of higher energy also being included in the analysis (Fig. 21). Once this fact is accepted, the problem becomes that of deciding optimal instrumental settings to determine the two isotopes.

The most obvious way to carry out an optimization is to maximize the counting efficiency for the lower energy isotope and then successively increase the upper window while

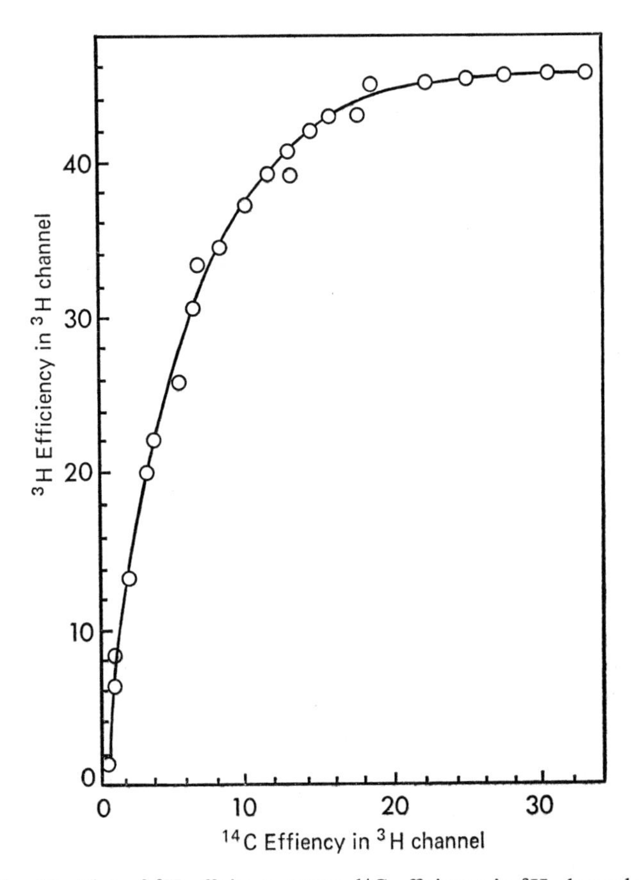

FIG. 22. Plot of ^{3}H efficiency versus ^{14}C efficiency in ^{3}H channel for air-quenched samples in toluene + PPO (Kobayashi and Maudsley, ref. 6).

noting the efficiencies for each isotope at each new discriminator setting. If ^{14}C and ^{3}H are the two isotopes a graph can be constructed of tritium efficiency versus ^{14}C efficiency in the tritium channel. A tangent to this curve gives the best efficiency ratio (Fig. 22).

A recent alternative is the use of an Engberg plot (see chapter 4 in ref. 6 of the Bibliography), where the counting efficiency for two isotopes is calculated in a fixed counting window with increasing gain. The results are plotted on a log–log basis as shown in Fig. 23. The best ratio is obtained

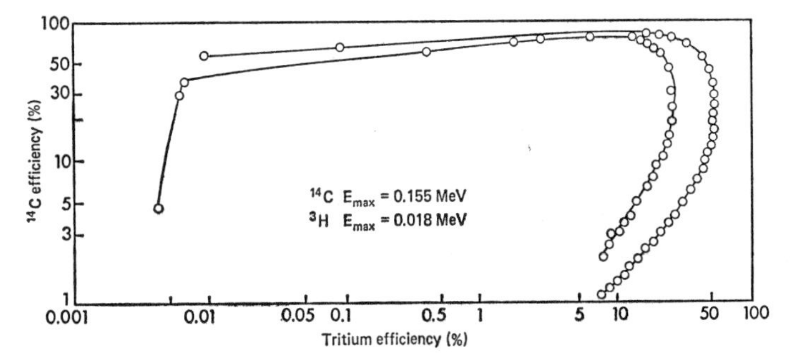

FIG. 23. Engberg plot for isotope pair ^{14}C and ^{3}H. Upper curve is derived from air-quenched samples in toluene + PPO; lower curve from acetone-quenched samples in toluene + PPO (Kobayashi and Maudsley, ref. 6).

by drawing a line at 45° to the end of the curve and the point at which this line deviates from the curve indicates the best counting parameters. Engberg plots can be applied in the presence of quenching conditions.

The contribution of each isotope to the total count can be calculated from simultaneous equations which yield $C = N_1/c_1$ and $H = (N_2 - Cc_2)/h_2$ where C is the number of disintegrations per minute of ^{14}C, H is the number of disintegrations per minute of tritium, c_1 is the ^{14}C efficiency in channel 1, c_2 is the ^{14}C efficiency in channel 2, h_2 is the ^{3}H efficiency in channel 2, N_1 is the net total counts observed in channel 1, and N_2 is the net total counts observed in channel 2. Dual-labelled counting is at non-balance point conditions and therefore small instrumental variations in amplifier gain etc. can cause serious errors. Obviously standardization in the normal manner at balance point is pointless and the use of the "spill-over" of the lower energy isotope into the higher channel is recommended as a routine measure of instrument stability (see also the Klein–Eisler[81] criterion mentioned on p. 76).

The simultaneous determination of two isotopes in heterogeneous samples is extremely difficult due to the inherent uncertainties in the determination of counting efficiencies. Better results are obtained by separate analysis for each isotope.

TABLE XV. Components of scintillator solutions studied by Birks and Poulis

Solvent	*Primary solutes*	*Secondary solutes*
Benzene	TP	PBBD
Toluene	PPO	POPOP
Xylene	BBOT	DMPOPOP
p-Xylene	PBO	BBO
Mesitylene	Butyl-PBD	bis-MSB
p-Dioxane†	PBD	α-NPO
	BIBUQ	DPHT

†With 100 g/l naphthalene.

When double-isotope counting is attempted under quenching conditions the quenched higher energy spectrum can be pushed almost entirely into the lower energy channel. In this case the automatic quench compensation (AQC) method has a special advantage.

All dual-labelled determinations at low count rates are troublesome and users must be aware that backgrounds are themselves liable to quenching. Indeed any method (single or double isotope) relying on a samples channels ratio technique may be imprecise when counts are close to background. Severe inconsistencies have been recorded when counting strongly quenched samples of ^3H in the presence of ^{14}C by a samples channels ratio method. These errors become apparent even with an observed ^3H count rate of just less than 2000/min.

TABLE XVI. Rankings of materials from Table XV relative to quench resistance

A. *Absence of quenching*

(i) *Order of solvent preference*
Toluene, *p*-xylene > xylene, mesitylene > benzene, dioxane+naphthalene

(ii) *Order of primary solute preference in toluene or* p-*xylene with solute concn in g/l in parentheses*
BIBUQ (24) > PBD (12); butyl-PBD (12); PBO (7·5) > BBOT (8) > PPO (7) > TP (7)

B. *Presence of impurity quencher* (10^{-2}M)†

(i) *Order of solvent preference*
Benzene, dioxane+naphthalene, > toluene, xylene > p-xylene, mesitylene

(ii) *Order of primary solute preference in toluene with solute concn in g/l in parentheses*
PBD (12); butyl-PBD (12); PBO (7·5) > BBOT (8) > PPO (6) > BIBUQ (15); TP (5)

C. *Secondary solutes*

Assessment of the merits of secondary solutes is not easy but a rough order, *in the absence of quenching*, is
PBBO > POPOP > bis-MSB > PBO > DMPOPOP > α-NPO > DPHT

† It is stressed that different behaviour with other impurity quenchers is anticipated.

G. Critical assessment of the liability of scintillator solutions to quenching

It was mentioned (p. 14) that an accurate appreciation of the performance of a scintillator solution is a complex concept affected by several parameters. One aspect which has not received detailed consideration until recently is the susceptibility of the solute and solvent to quenching. A full appraisal of this in relation to all possible cocktails is a long way from attainment, but Birks and Poulis[9] have started a programme of study. They are assessing the vulnerability to impurity quenching of some common solute+solvent combinations using a modern instrument with bialkali PM tubes.

Currently, carbon tetrachloride has been used to quench scintillations in mixtures drawn from the components listed in Table XV. Birks and Poulis use relative pulse height as the comparative parameter and all measurements were in glass vials. Their general conclusions are summarized in Table XVI.

Chapter 7

SOME GENERAL POINTS IN LIQUID SCINTILLATION PRACTICE

A. Instrument performance criteria[6]

For a single isotope in a counting window the simplest measurement of instrumental performance is provided by the parameter E^2/B, where E is the observed counting efficiency and B the background count. When unquenched samples are being measured E^2/B is determined in a differential counting mode by maximizing the count rate in a set window by use of gain controls (Fig. 24). This procedure is called "balance-point" counting and the balance-point is a function

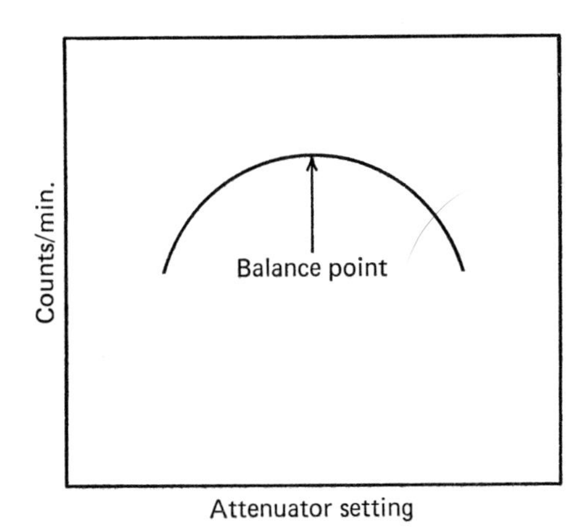

FIG. 24. Use of attenuator setting to achieve balance-point conditions.

of the β energy. In quenched samples E^2/B is not necessarily a good criterion as it is affected by the shape of the energy spectrum which will be distorted, and shifted, when quenched. For quenched samples it has been recommended that the ratio of count rate (CR) to background is maximized. This gives the best counting conditions provided that $CR/B \gg 1$.

The E^2/B assessment cannot be used for dual-label determinations as these are not carried out at balance point. Instrument performance for double isotope counting can be quantified by the performance number P, as defined by Klein and Eisler (1966).[81] Taking ^3H and ^{14}C as a suitable example P is given by

$$P = S\,(^3\text{H counting efficiency})\,(^{14}\text{C counting efficiency})/1000$$

S is a separation factor which is a function of R_1 and R_2 the channels ratios for ^3H and ^{14}C, respectively, such that

$$S = (R_2 + 0.1\,R_2R_1)\,(1 + 10\,R_2)/(R_2 + 10\,R_2R_1) \times (1 + 0.1\,R_2).$$

For careful work long-term instrumental stability is just as important as performance criteria.

B. Luminescence

The modern liquid scintillation spectrometer has been designed to meet stringent requirements for the measurement of small quantities of light. This means that the counter readily responds to light from sources other than the fluorescent emission of the scintillator. There are a number of other light sources inherent to liquid scintillation counting.

One such source is the plastic caps commonly provided by vial manufacturers. These can be activated by the ultraviolet components of both sunlight and strip-lighting. This can produce spurious high backgrounds and can be very noticeable in ambient temperature counting systems. It can be reduced by using a darkened counting laboratory and tungsten lighting in laboratories where samples are prepared. A period of "dark-adaption", i.e. storage in the absence of light, is advisable prior to counting. In refrigerated machines the problem is less noticeable but only because the half-life of the induced vial-cap fluorescence decay is increased.

The vial-caps are sold as "disposable" but users are tempted to recycle, after cleaning, as a simple economy. This practice also can induce unwanted luminescence the source of which probably is the "baking-on" of detergents. The same comment applies to the reusage of polythene counting vials and to a lesser extent to glass counting bottles. Even the use of a polythene scoop to weigh solutes has been reported as a source of fluorescence.[9]

The major source of unwanted light comes from the production of a chemiluminescent reaction in the sample. Light-producing chemical processes can arise in a variety of ways during sample preparation.

(i) From the presence of oxygen in the sample, particularly if produced by combustion. The presence of oxygen can lead to the production of a chemically reactive excited oxygen molecule capable of inducing chemiluminescence.

(ii) From the presence of peroxides either from (i) or as impurities.

(iii) In any alkaline sample, which means that peroxides as agencies for removing colour should be avoided if a basic solubilizer (e.g. NaOH, KOH, NCS, Hyamine, Soluene) is needed. (Note: Beckman Bio-Solv BBS-2 is an acid solubilizer).

(iv) From contamination of emulsifiers and solvents. Even toluene can contain peroxides, and dioxane is notorious for the presence of these impurities. Phenols and quinones are said to be impurities liable to produce chemiluminescence.

The cures for chemiluminescence are many and varied. The most obvious is avoidance by using combustion methods and the reader is reminded of previous comments on these methods relating to strongly coloured samples. The alternative to combustion for coloured samples is to bleach out the colour and then create sample homogenization with an alkaline solubilizer. This has been a common way of counting labelled urine, tissue and blood samples. It has been demonstrated that this method induces intense luminescence which can last for weeks.[7] This chemiluminescence can be reduced by the addition of acid to reduce pH to less than 7, but this does not restore normal backgrounds even after three hours and acid addition will increase quenching. Sometimes this limited correction is still acceptable but it is almost invariably good practice to combust highly coloured samples rather than use bleaching agents (e.g. hydrogen peroxide or benzoyl peroxide). It is true that oxygen may still be introduced by combustion, but flushing the scintillator with nitrogen or argon before counting reduces its deleterious effects.

Peroxide induced luminescence has been reduced by adding the enzyme catalase but this introduces a further quenching agency. Oxygen scavengers have been added to scintillator solutions and ascorbic acid or tin tetrachloride are suitable for reducing chemiluminescence, and the common usage of an antoxidant in dioxane has been mentioned previously (pp. 9, 26).

When luminescence is observed in counting ^{14}C, or isotopes of higher β energies, it can be combated by raising the threshold (lower discriminator level) to exclude pulses of chemiluminescence origin as these are usually of low intensity. This is a feasible method with tritium provided that the reduction of the instrumental counting efficiency to about one half of its optimal value is acceptable to the user.

Reduction in temperature of the counting chamber will slow down the chemical reaction producing the unwanted light and hence reduces observed chemiluminescence. This is unacceptable for precise work as the decay will still continue over long periods of time (Fig. 25) and the user experiences continuing unreliable and variable backgrounds. Indeed, it may be good practice to temporarily *increase* temperatures before counting to aid the chemiluminescence decay.

Lastly it must be mentioned that instruments are available fitted with photon monitoring which can distinguish between true sample counts and the single photon events occurring in a counting bottle caused by other light-producing processes.

C. The sample counting bottle (vial)[25,56,57]

There are two types of glass vial available for purchase. They are those with a low background made from glass with a lower potassium content (i.e. low in ^{40}K) and cheaper ones with a higher inherent background. Other vials made

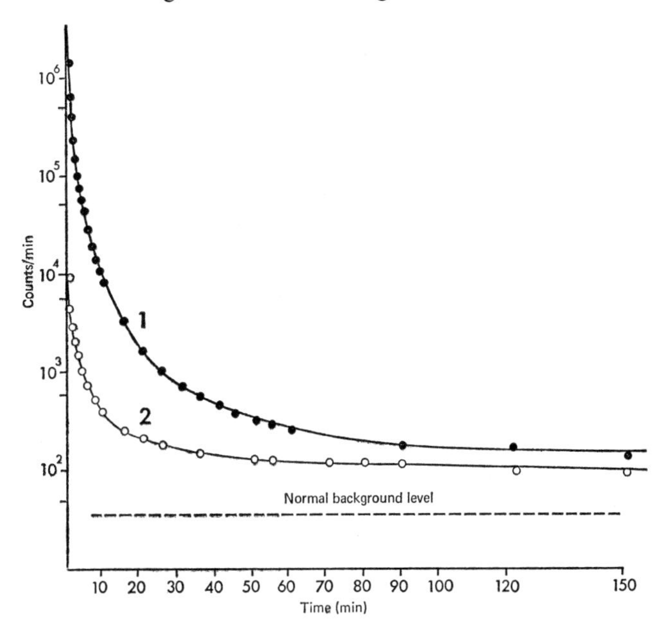

FIG. 25. Decay and duration of chemiluminescence reaction at 10 °C in two scintillation mixtures containing 1.0 ml Hyamine 10-X (counts on log-scale) (Kalbhen, ref. 7). 1. 15 ml Butler cocktail + 1 ml Hyamine; 2. 15 ml BBOT cocktail + 1 ml Hyamine.

from various polymeric materials are marketed and a comparison of the relative merits of all the types of vial in common use is in Table XVII.

When AESCR is the chosen quench correction method the use of polythene vials is to be avoided as the method is susceptible to the variations in vial material and to the permeation of solvent and solute through the vial walls. The use of glass vials is satisfactory for AESCR calibrations.

TABLE XVII. Comparison of liquid scintillation counting bottles

Material	*Inherent*† *background* (*counts/min*)	*Comments*
Glass with low potassium content	^{14}C 20 ^{3}H 16	Expensive—can be reused after careful cleaning
Glass	^{14}C 70 ^{3}H 60	Glass has a relatively poor transmittance in the scintillator emission range. Can be reused. Cheap
Nylon	^{14}C 16 ^{3}H 8	Low inherent background, some permeability to scintillation chemicals. Fluoresce‡ after exposure to u.v. light. Can require up to 4h dark adaption. Relatively poor transmittance in scintillator emission range. Not suitable for use with Instagel
Polythene§	Similar to nylon	Cheap and disposable but very permeable to toluene. Vials can distort and can pick up static charge which causes observable light flashes in the PM. Good transmittance to light of wavelength for scintillator emission
Teflon	^{3}H 8	Inert to scintillation chemicals, no induced phosphorescence. Reusable. Does not distort with time
Quartz	lowest of all	For very specialist low level counting (e.g. in dating). Cost too high for normal usage.

† Measured in the presence of scintillator solution. Only intended as a rough guide as they depend upon individual instrument and instrumental settings.

‡ Glass and polythene vials also fluoresce after exposure to u.v. light. The decay of that produced in glass is rapid (2 min) but that in polythene may persist over a longer period.

§ Polypropylene seems to be a better alternative to polythene.

It has been reported that glass vials are less prone to adsorb sample material onto their internal surface. The statement was made in relation to biological materials and is not true for inorganic isotopes. Siliconization of the internal surfaces of glass vials may be needed to prevent adsorption particularly when dealing with trace concentration of inorganic cations.

If vials are to be reused the normal cleaning procedure is: (i) rinse with solvent, (ii) soak in decontaminating solvent (e.g. Decon 75), preferably in an ultrasonic cleaning bath, (iii) rinse several times with distilled water, and (iv) dry in an air-oven. Cleaning of glass vials in an acid-cleaning bath leaches the glass surface and greatly increases the chance of contamination by adsorption.

D. Vial caps

A variety of caps are available. The standard cap contains an insert of cork with a tin foil liner to act as a light reflector to improve light collection. Users requiring an inert liner can purchase caps lined with a polythene disc. Polythene snap caps can be bought as very cheap disposable alternatives. This type of cap can be troublesome when used in an automatic counter as they may not be of the right dimensions to activate automatic sample changes which work via microswitches or "beam-interruption" devices. In most sophisticated instruments special caps are provided which act as "keys" to call up pre-programmed instrumental settings specific to the batch of samples to be counted. These are the so-called "multi-user" instruments.

E. Sample volume

In a modern instrument variation of sample volume has little effect on counting efficiency provided that the volume is in the range 6 ml to 15 ml. A "rule of thumb" guide is that a good instrument should show about a $\pm$ 2% variation in efficiency between 8 and 12 ml.

It has already been emphasized that methods of external standardization are *not* independent of sample volume.

F. Temperature control

Refrigerated counting compartments originally were built into counters to reduce thermal noise in the PM tubes. With modern bialkali tubes thermal noise is negligible below 40 °C, thus it seems that operation below room temperature is no longer needed. This is not necessarily true and Rapkin[39] has presented an excellent critical analysis of the question of temperature control. The factors considered and his conclusions, together with some other points culled from elsewhere in this book and repeated for convenience, can be summarized under the following sub-headings.

1. PM tube noise

Suitably matched bialkali tubes operated in coincidence can give ^{3}H counting efficiencies of 55 to 60% at ambient temperatures.

2. PM tube sensitivity

A constant operating temperature is not critical except when counting highly quenched sample or tritium in a dual-labelled experiment.

3. Sample solubility problems

Operation at ambient temperatures solely to increase the water "capacity" of a cocktail is not justified due to the increased quenching occurring with increased water content. In addition, increased sample solubility may be no advantage for the same reason. Ambient temperature operation may mean that TP, a cheap solute, can be used and that a higher concentration of primary solute can be used to diminish quenching.

4. Choice of solute and solvent

PPO and naphthalene have high solubilities in scintillator solvents even at 0 °C. TP and PBD have inferior solubilities in toluene at 0 °C and this restricts their effective use to room temperature operation. Dioxane freezes at 11.7 °C and requires the addition of an antifreeze (e.g. ethylene glycol) for low temperature counting. Controlled low temperature reduces the possibility of errors due to losses by evaporation.

5. Scintillator efficiency

This is improved by cooling as this reduces thermal (i.e. radiationless) dissipation of energy which competes with the solvent to solute energy transfer processes (see Fig. 1).

6. Quenching

As this is a diffusional process its effect will be reduced by cooling. Oxygen quenching will also be reduced by a reduction in temperature as this lowers the solubility of oxygen in the scintillator solution.

7. Chemiluminescence

The rate of a chemiluminescent reaction will be reduced by cooling. This may be an acceptable way to achieve a reasonably constant background to a high activity sample but can cause serious errors in low activity samples when the continuing slow fluorescent decay gives variable backgrounds. A low temperature decreases the possibility of the formation of the reactive oxygen species responsible for the prompting of chemiluminescent reactions.

8. External standardization

Accurate temperature control is advised but it is not necessary to refrigerate.

9. Emulsion counting

Temperature *control* is essential and best counting efficiencies are observed at about 0 °C.

10. General conclusion

It seems that good temperature control is sound experimental practice and that refrigeration has many advantages which will justify the additional expense of a machine with a cooled counting chamber. In laboratories where biological samples are analysed, a cooled chamber will be essential as the problems of highly quenched samples and heterogeneous sample preparation will be endemic. Accurate temperature control enables radioactive standards to be prepared by weighing rather than volume dispensing.

Chapter 8

SOME OTHER USES OF LIQUID SCINTILLATION COUNTERS

A. Suspended scintillators

Early attempts to count aqueous and alcoholic solutions of radioisotopes described the suspension of a solid scintillator in the solution. First experiments were to use plastic scintillators in the form of beads or bundles of filaments. These were inferior to the use, described by Steinberg,[93] of crystalline anthracene. This material was used in its "blue-violet fluorescent" grade and trace quantities of detergent were added to wet the crystals.[85] The method has a very low tritium efficiency ($\sim$0·5%) but figures of $\sim$20% for ^{14}C were recorded and other workable efficiencies for ^{45}Ca (93%), ^{131}I (46%) and ^{32}P (78%) were obtained. The technique is not affected greatly by pH and $H_3{}^{32}PO_4$ and $H_2{}^{35}SO_4$ have been counted in this way. Atmospheric ^{85}Kr has been estimated in a gas-tight vial fitted with scintillator shavings[92] (counting efficiency 94%).

An interesting variant on this approach was the development, by Heimbuch,[70] of an ion-exchange resin into which 9,10-diphenylanthracene or *p*-terphenyl can be copolymerized. The resulting beads can be used in both a cationic and anionic form even in concentrated acid or alkaline solutions. Heimbuch has used the "scintillating ion-exchange resin" to determine ^{35}S, ^{36}Cl, ^{63}Ni, $^{90}Sr/^{90}Y$, ^{131}I, ^{237}Np and ^{239}Pu.

B. Flow cells[6,47]

A logical extension of the suspended scintillator concept was the evolution of a cell, capable of accepting a reasonable flow of liquids, into which a solid scintillator could be packed. Early cells were designed employing thin plastic scintillators in the form of sheets, tubing or beads. These are of less utility than subsequent designs into which anthracene crystals are packed. Best counting efficiencies are gained by the continuous mixing of the liquid with a scintillator solution before passing it into a suitable cell positioned in the counting chamber. This method is destructive, susceptible to small changes in water content, and difficult to operate with aqueous solutions.

Despite the obvious drawbacks to flow cell work it has proved invaluable in amino-acid analyses, protein titrations and for continuous determinations of the isotopes ^{40}K, ^{45}Ca, ^{90}Sr/^{90}Y in acid and salt solutions. Another tempting use of flow cells is to use them to monitor effluent gases from a gas chromatograph.[30] A truly continuous flow cell for gases has not yet been constructed, but successful methods exist whereby volatile products separated by gas chromatography have been examined semi-continuously by (i) passage through scintillator tubing or anthracene crystals; (ii) successive fractionations into cartridges of silicon coated anthracene crystals or (iii) absorption into scintillator solutions.

C. Special vials

Ashcroft[48] has developed an annular counting vial in which the outer compartment is filled with a PPO+POPOP in toluene solution loaded with tetrabutyl tin. This outer annular ring is sealed. Samples of ^{125}I introduced into the centre of the vial can be counted with high efficiency. The tin acts as a focus to increase the absorption of the ^{125}I γ-rays thus increasing the possibility of activating the scintillator. A counting efficiency of 55% was recorded compared with one of 10% obtained when the ^{125}I sample was blended into a cocktail. Other γ emitters can be determined in the same way.

D. The analytical use of quenching

The modern liquid scintillation spectrometer is designed to be sensitive to quenching and it may be possible to use this sensitivity as a quantitative measure of quencher concentration. One publication (Zemskova *et al.*[96]) describes the study of the quenching of a toluene scintillator by iron and copper powders and suggests that concentrations of these quenchers may be determined to within $\sim 10^{-5}$g. Another application[8] uses the property of the spectral shift caused by controlled additions of chloroform as a means of identification for the isotopes ^{3}H, ^{14}C, ^{35}S, ^{36}Cl, ^{63}Ni and ^{185}W. They are identified in samples of biological and environmental origin. The shift in pulse-height of ^{14}C labelled formic acid, caused by glass spheres of varying size, has been suggested as a means of determining mean particle size distributions (Winkler and Vogt[95]). These novel uses of liquid scintillation phenomena may be the beginnings of an even wider use of the modern automatic instruments.

E. The analytical use of chemiluminescence[8,9]

The sensitivity of contemporary instruments to small amounts of light has prompted their use as quantitative light detectors. The availability of luminescence-producing enzymes has resulted in the development of the highly sensitive bio-assay of adenosine triphosphate (ATP) by measuring the light output from its reaction with the firefly enzyme luciferin-luciferase. The best analysis is obtained by counting out of coincidence in either an ambient temperature or a cooled spectrometer. The repeat count mode is employed, and a calibration plot of the logarithm of counts per unit time versus the logarithm of moles ATP is used (Fig. 26). The range of detection is between 10^{-7} and 10^{-14} mole ATP. The technique can be extended to a semi-automatic process and to analyse for ADP and AMP.

Other bioassays[9] are those of reduced nicotinamide adenine dinucleotide (NADH) and flavin mononucleotide (FMN) using the dehydrogenase + luciferase system derived from *Photobacterium fischeri*, a marine bacterium. The firefly enzyme can also be used to assay the enzyme ATP-sulphurylase and as a means of estimating ammonia[86] in cellular systems.

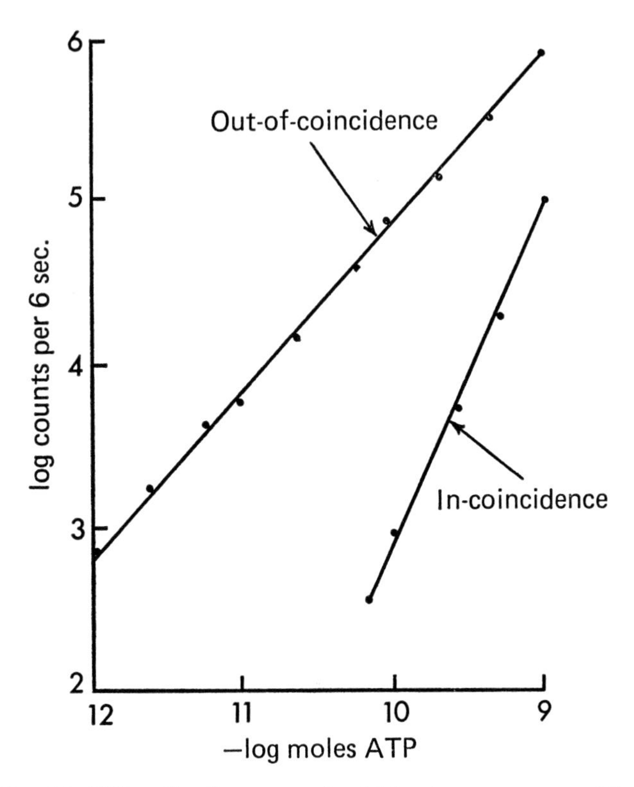

FIG. 26. ATP calibration curves for bioluminescence assay. (Stanley, ref. 8).

There seems no reason why the use of liquid scintillation equipment in this way should be restricted to bioluminescence assay. Kalbhen[7] has listed substances where chemiluminescence analysis is already used (Table XVIII) and presumably these analyses could be adapted for use with a scintillation spectrometer. He also points out that the luminol reaction is already used to promote chemiluminescence with organic peroxides, glucose, vitamin C, nitroaniline, resorcinol, pyrogallol, and other alcohols.

F. Čerenkov counting[6,8,32,42]

When a β particle of energy greater than about 300 keV travels through a medium of high dielectric constant (e.g. water) it produces local molecular polarization. These polarized molecules return to their ground state by the emission of electromagnetic radiation. When the velocity of the particle through the medium is greater than that of light through the medium the radiation is visible as u.v. light.

TABLE XVIII. Substances whose quantitative microassay is possible by use of a chemiluminescence reaction

Inorganic	*Organic*
Iron, cadmium, cobalt, copper, vanadium, cyanide, ozone, hydrogen peroxide	α-Amino-acids, oximes, organophosphorus compounds (Taburn, Sarin)

This phenomen is the Čerenkov effect and has been known since it was first observed by Mme Curie. The light is in the near u.v. and is detected by liquid scintillation spectrometers designed to cope with very low intensity light flashes. The detection is very similar to that of the scintillations caused by tritium and the instrumental settings for this isotope will also record Čerenkov radiation.

The number of photons emitted per Čerenkov event is proportional to the β energy causing it, so the method is most successful with hard β emitters and pulse height discrimination is possible.

Losses occur due to inefficient light collection, energy losses of the electrons in the medium and the broad energy distribution in the β spectrum.

Probably the figure of $\beta_{max} \approx 1$ is the limit of measurement for practical purposes. Parker and Elrick[6] list a series of experimentally determined Čerenkov counting efficiencies which illustrates this limitation (Table XIX).

Better efficiencies can be recorded with other PM tubes but they are very much a function of instrumental design. Čerenkov light is highly directional and of heterogeneous wavelength, therefore the sample volume must be carefully controlled. Attempts to optimize efficiencies have led various workers to operate their counters out of coincidence and this

TABLE XIX. Čerenkov counting efficiencies†

Radioisotope	*β Energy* E_{max} *(MeV)*	*Counting efficiency (%)*
^{204}Tl	0·77	1·3
^{36}Cl	0·71	2·3
^{198}Au	0·96	5·4
^{40}K	1·32	14
^{24}Na	1·39	18
^{32}P	1·71	25
$^{144}Ce/^{144}Pr$	2·98	54

†Measured for aqueous samples with an S11 PM tube. (Taken from Parker and Elrick, ref. 6 in Bibliography.)

certainly gives higher efficiencies. However, it is recommended that this be used with caution as backgrounds also increase and there is a possibility of multiple pulsing at the PM tubes. An alternative is to alter the directional property by using a wavelength shifter to absorb and then re-emit the Čerenkov energy isotropically in the visible region. A number of compounds have been used for this purpose including several naphthalene disulphonic acid derivatives (e.g. sodium potassium 2-naphthylamine 8-disulphonate). However, a recent survey by Ross[8] shows a clear preference for 4-methylumbelliferone.

The addition of a wavelength shifter detracts from the elegant simplicity of the direct non-destructive analysis of aqueous isotope solutions offered by the Čerenkov method but such an addition may still be worthwhile.

Because the production of Čerenkov light is a function of the relative velocities of light and the β particle through water, improvements[32] in detection efficiencies can be made by (i) increasing the refractive index or (ii) decreasing the specific gravity of the sample. The use of sugar to increase the refractive index adds about 7% to the counting efficiency of ^{40}K and ^{144}Ce/^{144}Pr and that of ^{36}Cl is improved by 10% by a change of 0·01 in refractive index. Another report[6] states that a specific gravity decrease of 7% improved the detection of ^{204}Tl by 10%. These effects are most significant with the isotopes of β energies close to the limit for the production of the Čerenkov effect. A preference for polythene vials is quoted and the ^{40}K content of glass will promote high Čerenkov backgrounds.

The method is not seriously affected by impurity quenching although organic solvents may themselves give weak Čerenkov emissions giving rise to spurious results. Chemiluminescence can occur but is not a major problem.

Colour quenching is operative and must be corrected for. Both samples channels ratio and AESCR are valid but now both are volume dependent[79]. Samples channels ratio is the advised method when the measured count rate is high. One proviso in the use of AESCR is that the energy of the γ standard must be high enough to create a Čerenkov effect (via its subsidiary extra-nuclear processes). This means that ^{137}Cs and ^{133}Ba standards are inadequate but ^{226}Ra is satisfactory.

Photon quenching of Čerenkov light has not been examined critically but there are several reports of Čerenkov determinations of labelled solids in suspension and there is no reason to doubt the validity of the results quoted. Such reports describe the assay of ^{36}Cl, ^{42}K and ^{86}Rb in suspended plant tissue.

As it may be construed from Table XIX this technique is invaluable for the estimation of ^{32}P in biological materials. ^{32}P assay in tissue slices, chromatographic media and bacteria now is carried out routinely by Čerenkov counting. Other noteworthy applications are its use to determine ^{42}K and ^{90}Sr in human urine and the detection of fall-out products in terrestrial waters. Iodine analyses in biological samples by neutron activation analysis followed by Čerenkov detection is a more esoteric use.

Continuous monitoring of radioactive effluents for ^{89}Sr, ^{90}Sr, ^{91}Y, ^{99}Mo, ^{140}Ba and ^{144}Pr is possible by this method and it has been suggested as a valuable adjunct to column chromatography.[9] Isotopes which have been determined by the Čerenkov effect are shown in Fig. 14.

The potential of this technique is considerable and attempts to lower the energy limit to that of ^{14}C by increasing refractive index—although unsuccessful—illustrate the great attraction the method has in its simplicity of sample preparation.

BIBLIOGRAPHY

The bibliography is arranged in four parts. The first lists text-books and conference reports. With each conference report a list of the chapters used for primary reference (and usually referred to in the text) is given. The second section lists the major review articles available in English. The third section quotes special publications distributed by commercial enterprise. The final part consists of the direct references to the original literature which have been used as source material. These are arranged in alphabetical order rather than the rough chronological order in the other sections.

A. Text-books and conference reports

(i) Text-books

1. Birks, J. B., "The Theory and Practice of Scintillation Counting". Pergamon Press, London (1964).
2. Schram, E., "Organic Scintillation Detectors". Elsevier, Amsterdam (1963).

(ii) Conference Reports

3. Bell, C. G. and Hayes, F. N., Eds., "Liquid Scintillation Counting". Pergamon Press, London (1958).
4. Proceedings of the University of New Mexico Conference on Organic Scintillation Detectors. USAEC, TID-7612 (1961).
5. Proceedings of the 1967 Summer School, Beckman Inst. Ltd. (1967). Source chapter: White, D. R., The evaluation of liquid scintillation mixtures of aqueous samples, p. 72.
6. Bransome, E. D., Ed., "The Current Status of Liquid Scintillation Counting". Grune & Stratton, New York (1970). Source chapters: Birks, J. B., Physics of the liquid scintillation process, p. 1; Kobayashi, Y. and Maudsley, D. V., Practical

aspects of double isotope counting, p. 76; MOGHISSI, A. A., Low-level liquid scintillation counting of α & β emitting nuclides, p. 86; SCHRAM, E., Flow monitoring of aqueous solutions containing weak β emitters, p. 95; PARKER, R. P. and ELRICK, R. H., Čerenkov counting as a means of assaying β emitting nuclides, p. 110; BRAY, G. A., Determination of radioactivity in aqueous samples, p. 170; GREEN, R. C., Heterogeneous systems-suspensions, p. 189; MAHIN, D. T. and LOFBERG, R. T., Determination of several isotopes in tissue by wet oxidation, p. 212; DAVIDSON, J. D., OLIVERIO, V. T. and PETERSON, J. I., Combustion of samples for liquid scintillation counting, p. 222; SNYDER, F., Liquid scintillation radioassay of thin-layer chromatograms, p. 248; GROWER, M. F. and BRANSOME, E. D., Liquid scintillation counting of macromolecules in acrylamide gels, p. 263; PENG, C. T., A review of methods of quench correction in liquid scintillation counting, p. 283; SPRATT, J. L., Off-line digital computation, p. 247.

7. DYER, A., Ed., "Liquid Scintillation Counting", Vol. 1. Heyden, London (1971). Source chapters: KALBHEN, D. A., Chemiluminescence as a problem and an analytical tool in liquid scintillation counting, p. 1; KRICHEVSKY, M. I. and MACLEAN, C. J., Optimization techniques for computer-aided quench correction, p. 55; BENAKIS, A., A new gelifying agent in liquid scintillation counting (Poly-Gel-B), p. 97.
8. HORROCKS, D. L. and PENG, C. T., Eds., "Organic scintillators and liquid scintillation counting". Academic Press, New York (1971). Source chapters: SCHRAM, E., *et al.*, Applications of scintillation counters to the assay of bioluminescence, p. 125; GUPTA, G. N., New method of micro-liquid scintillation counting in plastic minibags for ^{3}H, ^{14}C and ^{32}P, p. 747; ROSS, H. H., Performance parameters of selected wave-shifting compounds for Čerenkov counting, p. 757; GLASS, D. S., A versatile computer orientated liquid scintillation system using the double-ratio technique, p. 803; DUGAN, M. A. and ICE, R. D., Beta emitter identification by quench analysis, p. 1055.
9. CROOK, M. A., JOHNSON, P. and SCALES, B., Eds., "Liquid scintillation counting", Vol. 2, Heyden, London (1972). Source chapters: BIRKS, J. B. and POULIS, G. C., Liquid scintillators, p. 1; RAPKIN, E., History of the development of the modern liquid scintillation counter, p. 61; SCALES, B., Questions regarding the occurrence of unwanted luminescence in liquid

scintillation samples, p. 101; SCHRAM, E. and ROOSENS, H., Semi-automatic microtransfer and cell for the bioluminescence assay of ATP and reduced NAD with scintillation counters, p. 115; DYER, A., Methods of sample preparation of inorganic materials, including Čerenkov counting, p. 121; BURLEIGH, R., Liquid scintillation counting of low levels of ^{14}C for radiocarbon dating, p. 139; FOX, B. W., Sample preparation technique in biochemistry with particular reference to heterogeneous systems, p. 189; SPRATT, J. L., Acquisition and handling of liquid scintillation data, p. 245.

B. Review articles

10. DAVIDSON, J. D. and FEIGELSON, P., Practical aspects of internal sample liquid scintillation counting, *Int. J. Appl. Radiat. Isotopes* **2,** 1 (1957).
11. RAPKIN, E., Liquid scintillation counting 1957–1963—A review, *Int. J. Appl. Radiat. Isotopes* **15,** 69 (1964).
12. DYER, A., The development and scope of the technique of liquid scintillation counting, *Bio-Med. Eng.* 461 (1966).
13. FOX, B. W., Sample preparation techniques for scintillation counting, *Lab. Pract.* **17,** 595 (1968).
14. PARMENTIER, J. H. and TEN HAAF, F. E. L., Developments in liquid scintillation counting since 1963, *Int. J. Appl. Radiat. Isotopes* **20,** 305 (1969).
15. GIBSON, J. A. B. and LALLY, A. E., Liquid scintillation as an analytical tool, *Analyst* **96,** 681 (1971).
16. PRICE, L. W., Practical course in liquid scintillation counting, part I, Principle and chemistry, *Lab. Pract.* **22,** 27 (1973); part II, Preparing of samples—I, *Lab. Pract.* **22,** 110 (1973); part III, Preparing of samples—II, *Lab. Pract.* **22,** 181 (1973); part IV, The practical counter and quench correction, *Lab. Pract.* **22,** 277 (1973); part V, Computing and calculating techniques. *Lab. Pract.* **22,** 352 (1973).

C. Special publications

17. HAYES, F. N., Solutes and solvents for liquid scintillation counting, Packard Technical Bulletin, No. 1 (revised 1963).
18. HORROCKS, D. L., Liquid scintillation counting of inorganic radioactive nuclides, Packard Technical Bulletin, No. 2 (revised 1962).

19. RAPKIN, E., Hydroxide of Hyamine 10-X, Packard Technical Bulletin, No. 3 (revised 1970).
20. DAVIDSON, E. A., Techniques for paper-strip counting in a scintillation spectrometer, Packard Technical Bulletin, No. 4 (revised 1970).
21. RAPKIN, E., Liquid scintillation measurement of radioactivity in heterogeneous systems, Packard Technical Bulletin, No. 5 (revised 1963).
22. RAPKIN, E., The determination of radioactivity in aqueous solution, Packard Technical Bulletin, No. 6 (revised 1963).
23. RAPKIN, E., Measurement of $^{14}CO_2$ by scintillation techniques, Packard Technical Bulletin, No. 7 (1962).
24. GIBBS, J. A., Liquid scintillation counting of natural radiocarbon, Packard Technical Bulletin, No. 8 (revised 1970).
25. RAPKIN, E. and GIBBS, J. A., Polyethylene containers for liquid scintillation spectrometry, Packard Technical Bulletin, No. 9 (1965).
26. JEFFAY, H., Oxidation techniques of preparation of liquid scintillation samples, Packard Technical Bulletin, No. 10 (1962).
27. RAPKIN, E., Liquid scintillation counting with suspended scintillators, Packard Technical Bulletin, No. 11 (1963).
28. TAMERS, M. A., Liquid scintillation of low-level tritium, Packard Technical Bulletin, No. 12 (1964).
29. SCHILLING, R. F., Estimation of radioactive vitamin B-12 absorption with a large volume liquid scintillation detector, Packard Technical Bulletin, No. 13 (1964).
30. KARMEN, A., Measurement of radioactivity in the effluent of a gas chromatography column, Packard Technical Bulletin, No. 14 (1965).
31. HERBERG, R. J., Channels ratio method of quench correction in liquid scintillation counting, Packard Technical Bulletin, No. 15 (1965).
32. HABERER, K., Measurement of β activities in aqueous samples utilizing Čerenkov radiation, Packard Technical Bulletin, No. 16 (1966).
33. DE SOMBRE, E. R. and JENSEN, E. V., The digital computer as a laboratory accessory, Packard Technical Bulletin, No. 17 (1967).
34. "Preparation of samples for liquid scintillation counting" Nuclear Chicago (1967).
35. "Liquid Scintillation Counting". 3 Edn., Nuclear Chicago (1969).

36. NODINE, J. H., Digital computers in nuclear radiation counting, The Nucleus, No. 19, Nuclear Chicago (1965).
37. RAGLAND, J. B., Combustion of samples for liquid scintillation counting, The Nucleus, No. 20, Nuclear Chicago (1966).
38. TURNER, J. C., Sample preparation for liquid scintillation counting, Review No. 6, Radiochemical Centre, Amersham (revised 1971).
39. RAPKIN, E., Temperature control in liquid scintillation counting, Digitechniques, No. 1, Intertechnique.
40. RAPKIN, E., Sample preparation for liquid scintillation counting, Part I, solubilization, Digitechniques, No. 2, Intertechnique.
41. RAPKIN, E., Sample preparation for liquid scintillation counting, Part II, solvents and scintillators, Digitechniques, No. 3, Intertechnique.
42. PARKER, R. P., Čerenkov counting, Digitechniques, No. 4, Intertechnique.
43. RAPKIN, E., Gel and emulsion counting of aqueous solutions, Digitechniques, No. 5, Intertechnique.
44. BIRKS, J. B., An introduction to liquid scintillation counting, Koch-Light (1972).
45. BIRKS, J. B., Solutes and solvents for liquid scintillation counting, Koch-Light (1972).
46. NICOLL, D. R. and EWER, M. J. C., Liquid scintillation sample preparation techniques for organic materials, Nuclear Enterprises (1971).
47. VERHASSEL, J. P., Review of continuous flow techniques, from Nucleonics Congress Frankfurt 1972, ICN Tracerlab.

D. Principal source references from the original literature

48. ASHCROFT, J., Gamma counting of ^{125}I using a metal-loaded liquid scintillator, *Anal. Biochem.* **37**, 268 (1970).
49. BAILLIE, L. A., Determination of liquid scintillation counting efficiencies by pulse height shift, *Int. J. Appl. Radiat. Isotopes* **8**, 1 (1960).
50. BENSON, R. H., Limitations of tritium measurements by liquid scintillation counting of emulsions, *Anal. Chem.* **38**, 1353 (1966).
51. BRAY, G. A., A simple efficient liquid scintillator for counting aqueous solutions in a liquid scintillation spectrometer, *Anal. Biochem.* **1**, 279 (1960).

52. BUSH, E. T., General applicability of the channels ratio method of measuring liquid scintillation counting efficiencies, *Anal. Chem.* **35**, 1024 (1963).
53. BUSH, E. T. and HANSEN, D. L., Improvement of liquid scintillation efficiencies by optimization of scintillator composition. Relative efficiencies of three new fluors, "Radioisotope Sample Measurement Techniques in Medicine and Biology". IAEA, Vienna (1965).
54. BUSH, E. T., Relative efficiency of a new liquid scintillation fluor *p*-bis (σ-methyl styryl) benzene, *Anal. Chem.* **38,** 1241 (1966).
55. BUSH, E. T., A double-ratio technique as an aid to selection of sample preparation procedure in liquid scintillation counting, *Int. J. Appl. Radiat. Isotopes* **19**, 447 (1968).
56. BUTTERFIELD, D. and MCDONALD, R. T., Comparison between materials used for liquid scintillation counting vials, *Int. J. Appl. Radiat. Isotopes* **23,** 249 (1972).
57. CALF, G. E., Teflon vials for liquid scintillation counting of tritium samples, *Int. J. Appl. Radiat. Isotopes* **20,** 611 (1969).
58. CARROLL, C. O. and HOUSER, T. J., Liquid scintillation counting: data analysis and computers, *Int. J. Appl. Radiat. Isotopes* **21,** 261 (1970).
59. FOX, B. W., The application of Triton X-100 colloid scintillation counting in biochemistry, *Int. J. Appl. Radiat. Isotopes* **19,** 717 (1968).
60. GIBSON, J. A. B. and GALE, H. J., A fundamental approach to quenching in liquid scintillators, *Int. J. Appl. Radiat. Isotopes* **18,** 681 (1967).
61. GIBSON, J. A. B. and MARSHALL, M., Counting efficiency of ^{55}Fe and other electron capture nuclides by liquid scintillation counting, *Int. J. Appl. Radiat. Isotopes* **23,** 321 (1972).
62. GILL, D. M., Liquid scintillation counting of tritiated compounds supported by solid filters, *Int. J. Appl. Radiat. Isotopes* **18,** 393 (1967).
63. GLASS, D. S., Automatic quench correction by channels ratio for ^{14}C and ^{3}H using a three channel liquid scintillation counter, *Int. J. Appl. Radiat. Isotopes* **21,** 531 (1970).
64. GOMEZ, E., *et al.*, Nitriles as scintillation solvents. Benzo- and acetonitrile, *Int. J. Appl. Radiat. Isotopes* **22,** 243 (1971).
65. GORDON, B. E. and CURTIS, R. M., The anti-quench shift in liquid scintillation counting, *Anal. Chem.* **40,** 1486 (1968).

66. HANSEN, D. L. and BUSH, E. T., Improved solubilization procedures for liquid scintillation counting of biological materials, *Anal. Biochem.* **18,** 320 (1967).
67. HARDCASTLE, J. E., *et al.*, Simultaneous assay of ^{45}Ca and ^{89}Sr in double isotope biological samples by liquid scintillation counting, Anal. Biochem. **46,** 534 (1972).
68. HAYES, F. N. and GOULD, R. G., Liquid scintillation counting of tritium labelled water and organic compounds, *Science* **117,** 480 (1953).
69. HAYES, F. N., *et al.*, Importance of solvent in liquid scintillators *Nucleonics* **13,** 46 (1955).
70. HEIMBUCH, A. H., *et al.*, The assay of α and β emitters with scintillating ion-exchange resins, IAEA, SM 61/65 (Conf. 650 507–24) (1965); *Nucl. Sci. Abstr.* **19,** 32098.
71. HIGASHIMURA, T., *et al.*, External standard method for the determination of the efficiency in liquid scintillation counting, *Int. J. Appl. Radiat. Isotopes* **13,** 308 (1962).
72. HENDEE, W. E., *et al.*, Tritiated toluene, tritiated water and tritiated hexadecane as internal standards for toluene and dioxane based liquid scintillators, *Int. J. Appl. Radiat. Isotopes* **23,** 90 (1972).
73. HORROCKS, D. L. and STUDIER, M. H., Determination of absolute disintegration rates of low energy β emitters in a liquid scintillation spectrometer, *Anal. Chem.* **33,** 615 (1961).
74. HORROCKS, D. L. and STUDIER, M. H., Determination of radioactive noble gases by liquid scintillation counting, *Anal. Chem.* **36,** 2077 (1964).
75. HORROCKS, D. L., Liquid scintillation methods for the calculation of the average energy to produce one photoelectron, *Nucl. Instr. Methods.* **27,** 253 (1964).
76. HORROCKS, D. L., Pulse height energy relationships of a liquid scintillator for electrons of energy < 100 KeV, *Nucl. Instr. Methods* **30,** 157 (1964).
77. HORROCKS, D. L., α particle energy resolution in a liquid scintillator, *Rev. Sci. Instr.* **35,** 334 (1964).
78. IORGULESCU, A., Use of a liquid scintillator to detect a gaseous capture emitter ^{37}Ar, *Compt. Rend., Ser. B* **273,** 278 (1971).
79. KAMP, A. J. and BLANCHARD, F. A., Quench correction in Čerenkov counting channels ratio and external standard channels ratio, *Anal. Biochem.* **44,** 369 (1971).
80. KINARD, F. E., Liquid scintillator for the analysis of tritium in water, *Rev. Sci. Instr.* **28,** 293 (1957).

81. Klein, P. D. and Eisler, W. J., An improved description of separation and performance capabilities of liquid scintillation counters used in dual isotope studies, *Anal. Chem.* **38,** 1453 (1966).
82. Krichevsky, M. I., *et al.*, Computer aided single or dual isotope channels ratio quench correction in liquid scintillation counting, *Anal. Biochem.* **27,** 442 (1968).
83. Lieberman, R. and Moghissi, A. A., Low-level counting by liquid scintillation. II-Application of emulsions to tritium counting, *Int. J. Appl. Radiat. Isotopes* **21,** 319 (1970).
84. Meade, R. C. and Stiglitz, R. A., Improved solvent systems for liquid scintillation counting of body fluids and tissues, *Int. J. Appl. Radiat. Isotopes* **13,** 11 (1962).
85. Myers, L. S. and Brush, A. H., Counting of α and β radiation in aqueous solutions by the detergent-anthracene scintillation method, *Anal. Chem.* **34,** 342 (1962).
86. Nicholas, D. T. D. and Clarke, G. R., Bio-luminescent method for the determination of microquantities of ammonia in a liquid scintillation spectrometer, *Anal. Biochem.* **42,** 560 (1971).
87. Patterson, M. S. and Greene, R. C., Measurement of low energy β emitters by liquid scintillation counting of emulsions, *Anal. Chem.* **37,** 854 (1965).
88. Paus, P. N., Liquid scintillation counting of RNA, a simple procedure for extraction from sucrose gradients, *Anal. Biochem.* **38,** 364 (1970).
89. Polesky, H. F. and Seligson, D., Application of liquid scintillation counting to ^{131}I measurements, *Anal. Biochem.* **10,** 347 (1965).
90. Rapkin, E. and Reich, A., Automatic combustion for routine liquid scintillation sample preparation, *Intern. Lab.* 31 (1972).
91. Rogers, A. W. and Moran, J. F., Evaluation of quench correction in liquid scintillation counting by internal, automatic external, and channels ratio standardization methods, *Anal. Biochem.* **16,** 206 (1966).
92. Sax, N. I., *et al.*, Modified scintillation counting technique for the determination of low-level ^{85}Kr, *Anal. Chem.* **40,** 1915 (1968).
93. Steinberg, D., Radioassay of aqueous solutions in the liquid scintillation spectrometer, *Anal. Biochem.* **1,** 23 (1960).
94. Van der Laarse, J. D., Experience with the emulsion counting of tritium, *Int. J. Appl. Radiat. Isotopes* **18,** 485 (1967).

95. WINKLER, K. and VOGT, H., Possibility to determine mean particle size distribution through quenching effects, *Isotopenpraxis* **7**, 148 (1971).
96. ZEMSKOVA, I. E., *et al.*, Mechanical quenching of the light of liquid scintillation counters, *Instrum. Exp. Tech. USSR* (Engl. Transl.). No. 1, p. 90 (1970).

APPENDIX

TABLE XX. Cocktails selected from the literature and their suggested use (see also Table VI)

Sample	*Cocktail*
Homogeneous sample in refrigerated systems	(i) 4 g/l PPO + 40 mg/l POPOP (or DMPOPOP) in toluene (ii) 8 g/l butyl-PBD in toluene
Slightly quenched sample in modern ambient counter	(i) 4 g/l PPO in toluene (ii) 6 g/l butyl-PBD in toluene
Quenched samples in modern counter	8 g/l PPO + 60 mg/l DMPOPOP in toluene
Tritiated water	(i) 4 g/l PPO + 20 mg/l POPOP in 10% methanol, 2% ethylene glycol, 6% naphthalene made up to 1 l with dioxane (ii) 2.75 : 1 *p*-xylene, Triton N-100 with 7 g/l PPO and 1.5 g/l bis-MSB
Aqueous salt solutions and some biological samples (e.g. urine)	(i) as for tritiated water cocktail (i) (ii) Triton X-100 in toluene (iii) Triton X-114 in xylene + naphthalene (See Greene, ref. 8 Chapter 19)
TLC spots	(i) 15 parts dioxane : 3 parts water with 7 g/l PPO and 30 mg/l DMPOPOP (labelled molecule stripped from support) (ii) Cab-O-Sil or HF/Triton X-100 in toluene (Labelled material remaining on support or when coloured solutions are present)

Several firms sell pre-prepared scintillator solutions or mixture of solutes e.g. Fluoraliy (Beckman), Perma Fluor (Packard), Scintisol, Scintimix, KL 359, 360, 371, 372 (Koch-Light), NE 216, 233 (Nuclear Enterprises).

TABLE XXI. List of commercially avaiable gels, gelling agents, solubilizers etc. and suggested uses according to sales literature

Product	*Chemical type (where known)*	*Suggested Uses*
NCS (Nuclear Chicago)	Quaternary ammonium base in toluene	Use with PPO and POPOP (but not butyl-PBD) for biological fluids, blood, whole tissues, tissue homogenates, CO_2, polyacrylamide gels, dry samples (e.g. collagen, RNA), aqueous inorganic salt and acid solutions
PCS (Nuclear Chicago)	Sold with premixed fluor	Aqueous solutions of inorganic and organic samples
Soluene-100 (Packard)	Quaternary ammonium base	Use in toluene with PPO and DMPOPOP *or* dioxane scintillator. For wet and dry tissue, aminoacids, proteins and plasma
Soluene-350 (Packard)	Quaternary ammonium base (Improved version of Soluene-100)	For whole tissue, homogenates, faeces, proteins, protein colloids, inorganic salts, serum, plasma, urine, whole blood, water soluble polysaccharides, aminoacids, nucleic acids, lipids, steroids, acrylamide discs, bacteria and cells, filter discs
Carbo-sorb (Packard)		High capacity for CO_2
Dimilume-30 (Packard)		Inhibits chemiluminescence
Insta-gel (Packard)	Solvent + solute + nonionic surfactant	High water capacity ($\rightarrow 50\%$), aqueous solutions of salts and sugars, serum, plasma, urine, whole blood,* erythrocytes,* homogenized tissue, protein precipitates, TLC spots, polyacrylamide gels (*use with Soluene-350)

TABLE XXI.—contd.

Product	*Chemical type (where known)*	*Suggested Uses*
Monophase-40 (Packard)		Specially designed fluor for use with aqueous solutions
KL-368 (Koch-Light)	Dioxane based cocktail with silica	Use with KL-35A and KL-358 (solutes from Koch-Light) for aqueous solutions
Unisolve 1 (Koch-Light)	Based on a pure aromatic hydrocarbon. Contains emulsifier but no dioxane, alcohol or glycol	Aqueous solutions of salts and sugars. Barium carbonate, magnesium carbonate, silica, alumina. Urine, milk, serum, plasma and other organic materials. Biological samples counted in Hyamine hydroxide with aqueous hydrochloric acid
NE-520 (Nuclear Enterprises)	Solubilizer	Takes in water, TCA, perchloric acid, urea, urine, sodium chloride, blood, plasma, aminoacids, proteins, tissue digests, carbohydrates
NE-220, NE-240, NE-250	Dioxane based cocktails	For aqueous samples
NE-221	Gel	For insoluble samples and aqueous solutions
NE-260 (Nuclear Enterprises)	Micellar scintillator	Multipurpose, takes in 1 ml of water, plasma, urine, per 10 ml scintillator
Triton surfactants (Rohm-Haas)	Detergent	Creating emulsions (and colloids) for aqueous solutions, suspended solids etc. (Types: Triton X-100, X-102, X-45, X-114, N-101)
Hyamine 10X (Rohm-Haas)	Quaternary ammonium compound available as chloride or hydroxide	Solubilizer for wide variety of biological samples and for CO_2

TABLE XXI.—contd.

Product	*Chemical type (where known)*	*Suggested Uses*
Poly-Gel B (see ref. 7)	Polyolefinic resin	Gel formed by heating with PPO (or butyl PBD) in toluene at 60–70 °C. Useful for biological solids, plant materials, TLC supports (dry material only)
Cab-O-Sil (Cabot Carbon Ltd.)	"Fumed" silica	Forms thixotropic gel when used as 4% (w/v) with toluene scintillators. Takes up insoluble materials
Bio-solv solubilizers (Beckman-RIIC)	Are not emulsions or dispersing systems	Use with toluene, xylene or benzene at ambient or refrigerated temperatures BBS-2, for alkaline tissue preparations, BBS-3 for aqueous, urine and plasma samples
Aquasol (New England Nuclear)	Xylene based (including fluors)	Acrylamide gels, proteins, lipids on TLC plates, biological tissue, urine, serum, inulin, sodium citrate buffers, inorganic salt solutions
Protosol (New England Nuclear)	Quaternary ammonium hydroxide solution	Lung, liver, kidney, spleen, heart, muscle, serum and plasma. (Tolerates 40% water at 20 °C, 25% at 5 °C). Also polyacrylamide gels and CO_2.

INDEX